THE FRIENDSHIP OF CANNOCK CHASE

British Library Cataloguing-in-Publication Data
A catalogue record for this book is available from the
British Library

Some places mentioned in the Book

The figures denote approximately the number of miles

ABBOTS BROMLEY.
6 from Rugeley.
BEAUDESERT PARK.
1 from Moor's Gorse.
2 ,, Longdon.
3 ,, Rugeley.
2½ ,, Hednesford.
5 ,, Lichfield.
BLITHEFIELD.
4½ from Rugeley.
3 ,, Abbots Bromley.
BROCTON.
4 from Pottal Pool.
1 ,, Milford.
CANK THORN.
3½ from Cannock.
¾ ,, Pye Green.
1 ,, Pottal Pool.
CANNOCK.
9 from Wolverhampton.
9 ,, Stafford.
8 ,, Walsall.
CASTLE RING.
3 from Hednesford.
1 ,, Beaudesert Hall.
COLWICH CHURCH.
3 from Rugeley.
3½ ,, Milford.
CONGREVE.
1½ from Penkridge.
DICK SLEE'S CAVE.
1 from Seven Springs (Colwich).
EDIAL HOUSE .
6 from Cannock.
2 ,, Lichfield.
ESSEX BRIDGE.
2 from Milford.
FAREWELL CHURCH.
2 from Castle Ring.
FOUR CROSSES.
2 from Cannock.
2¾ ,, Gailey.
2¾ ,, Four Ashes.
HATHERTON HALL.
1½ from Cannock.
HEDNESFORD.
2 from Cannock.
5 ,, Rugeley.
HILTON PARK.
4 from Cannock.
5 ,, Wolverhampton.
LONGDON.
3½ from Rugeley.
4 ,, Lichfield.
MARQUIS'S DRIVE.
5 in length from beginning at Beaudesert Park to Sherbrooke Valley end.
MESSINES MODEL.
1 from Brocton.
3 ,, Military Cemetery.
MILFORD.
3 from Stafford.
7½ ,, Cannock.
5½ ,, Rugeley.
MILITARY CEMETERY.
3¼ from Cannock
¾ ,, Pye Green.
1 ,, Pottal Pool.
MOOR'S GORSE.
2½ from Hednesford.
2½ ,, Military Cemetery.
2½ ,, Rugeley.
NUN'S WELL.
3 from Hednesford.
½ ,, Castle Ring.
OAKEDGE PARK.
4 from Military Cemetery.
1 ,, Wolseley Bridge.
2½ ,, Milford.
PENKRIDGE.
4½ from Cannock.
3 ,, Pottal Pool.
PILLATON HALL.
3¾ from Cannock.
2 ,, Penkridge.
POTTAL POOL.
3 from Cannock.
3 ,, Penkridge.
1 ,, Military Cemetery.
PYE GREEN.
2¾ from Cannock.
1½ ,, Hednesford.
RADMORE.
½ from Castle Ring.
RAWNPIKE OAK.
2½ from Hednesford.
½ ,, Castle Ring.
SEVEN SPRINGS.
¾ from Colwich.
1¼ ,, Wolseley Bridge.
SHARESHILL CHURCH
3 from Cannock.
6 ,, Wolverhampton.
SHOAL HILL.
1 from Cannock.
SHERBROOKE VALLEY.
4½ walk from Military Cemetery through the Valley to emerge on main road from Milford to Rugeley at a point 1½ from former and 4 from latter.
SHUGBOROUGH HALL
2 from Milford.
SOUTH STREET.
4½ walk from Military Cemetery to junction of Stafford-Rugeley main road near Wolseley Bridge.
STILE COP.
2 from Rugeley.
4 ,, Hednesford.
TEDDESLEY HALL.
4½ from Cannock via Shoal Hill.
1¾ ,, Penkridge.
5 ,, Stafford.
WATLING STREET in Chase area.
8 in extent from Gailey to Brownhills.
WEDGES MILLS.
1 from Cannock.
WOLSELEY BRIDGE and
WOLSELEY HALL.
2 from Rugeley.
3½ ,, Milford.
WYRLEY GROVE.
1½ from Norton Canes.
3½ ,, Cannock.

"PITMAN" (M. WRIGHT, J.P.).

THE FRIENDSHIP OF CANNOCK CHASE

by

"PITMAN"

(M. WRIGHT, J.P.)

Author of "The Best of Cannock Chase"

WITH A FOREWORD BY

LORD HATHERTON

FOREWORD

HATHERTON HALL,
CANNOCK.

It is a goodly thing in man that he should revere the soil from which he and his have sprung, and so it is with considerable pleasure that I respond to the request for a foreword to this book on Cannock Chase.

My ancestors always wrote of what we call the Chase as the "Hills" and, since for over five hundred years we have dwelt on its borders, I can truthfully say, like the Scotsman of the Highlands, that the breath of the "Hills" is in my blood.

In the fifteenth century when the King hunted on the Chase a Littleton was "Equitator" on "Cank" Chase, a position which corresponds, I imagine, to the office of Head Keeper. From that particular Official derive the Stag's Head and Bugle Horn which form our family crest.

The history of the Chase is a long and noble one, and from my earliest childhood I have listened to stories told by my forbears about the glories it offered for hunting and shooting and the wealth of grouse and blackgame (exterminated, alas, by the camps in the war) which it offered its guests.

I can remember as a boy hearing stories of riding and driving when the barouche would sink in the bogs and have to be dug out, when no roads, as we know them now, existed, and when modern means of transport were only an idle dream.

The Chase has done many services for man in its long history, but perhaps none more valuable than in our own time when it became one vast camp through which two million soldiers passed during the Great War. And now, once more amidst the blessings of Peace, its stimulating air and noble vistas are free for high and low alike to enjoy.

It is with genuine pleasure that I welcome in this book a tribute both to its enduring beauties and to the pageantry of its past.

HATHERTON.

October, 1935.

CONTENTS

	Page
FOREWORD	
THE FRIENDSHIP OF CANNOCK CHASE	13
JIM AND JOE SEE WONDERS	16
ALONG THE BOUNDS OF ANCIENT CANNOCK FOREST	21
SILVER BIRCHES AND A LOVELY SKY	25
BOGS TURNED INTO FERTILE LANDS	28
TALKING OF PLATO BENEATH A COPPER BEECH	32
STANLEY'S STORY OF THE MISSING CANDLE	37
CANNOCK CHASE DEER FOR A THOUSAND YEARS	44
WHEN PASTOR JEROME CAME FROM APPLEDORE IN DEVON	49
MACAULAY'S DAY OF PRAISE ON CANNOCK CHASE	53
THE BLUE PUMP AND THE CATHEDRAL	58
LITTLETON'S KEEN EYE FOR BEAUTY	61
I RAISE MY HAT ON MILFORD HILLS	66
A JUBILEE AND A DISTANT VOICE	70
HUNTSMEN AT HEDNESFORD FINGER POST	74
SPECTRAL SIR HUGH RECLAIMS HIS GOBLET SKULL	79
DIM CENTURIES IN COURT BANK COVERT	82
AMONG THE BOOKS AT WYRLEY GROVE	87
JOSEPH THE DREAMER	92
JAMES TURNER TELLS OF THE "POCKET VENUS"	96
"HALF-SOAKED DICK" ON MUSIC	101
A JUNE DAY ON ESSEX BRIDGE	106
ALL ARTS COMBINE IN TRENT-SIDE ARCADY	112
THE CEREMONY OF THE KEY	116
ON CANNOCK PARISH CHURCH TOWER	121
THE BLUEBELLS RETURN TO LONGDON WOODS	126
PATSY THE KING OF CANNOCK CHASE	129
THROUGH SPRING SLADE ON AN AUTUMN DAY	135
A GARDEN OF DELIGHT AT LONGDON GREEN	139
A POETESS AND THE BESOM CLUB	144
COME AND SEE CANNOCK CHASE	147
AN UNSOLVED CROSS KEYS MYSTERY	150
JACK OF HILTON EMERGES FROM HIS GREEN BAG	153
THE DEERS' DEAD HEAT DOWN WOODLAND TRACK	159
ON THE BAGOT LANDS FOR SIX CENTURIES	162
WHEN JOHN ORDERED OUT THE DUCHESS	167
THE LADY WITH THE DUSTER	172
JIMMY TELLS HOW THE COAL WAS FOUND	176
TUDOR ROSES GATHERED ON CANNOCK CHASE	181
MARY, HILDA, AND MADELINE	185
THE LAST OF THE FLEUR DE LYS	189
OLD SPORTSMEN WITH LONG MEMORIES	192
CANNOCK'S HARVEST OF SPARROWS' HEADS	197
IN FOOTSTEPS OF ANCIENTS IN BEAUDESERT PARK	201
THE VILLAGE BLACKSMITH WAS FIRED FOR SCIATICA	205
THE RIVER GOD OF CANNOCK CHASE	208
SID'S ENGLISH OF THE SONOROUS ROLL	212
POETRY AND PROSE IN BLITHEFIELD COURTYARD	216
OLD TIMOTHY WAS SURE ABOUT GENTLE SHAW	221
FAREWELL TO BEAUDESERT HALL	224
AN ADDENDUM OF SUGGESTED ROUTES	228

LIST OF ILLUSTRATIONS

"PITMAN" (M. WRIGHT, J.P.) - - - - *Frontispiece*

MR. ALBERT STANLEY, M.P. - - - *Facing page* 37

HATHERTON HALL - - - - - ,, 61

THE ESSEX BRIDGE - - - - - ,, 107

THE ANSON MONUMENT, SHUGBOROUGH PARK ,, 113

CANNOCK CHASE, FROM PYE GREEN - - ,, 135

BIRCHES VALLEY - - - - - ,, 187

WINDOW OF BANQUETING HALL, BEAUDESERT ,, 225

THE FRIENDSHIP OF CANNOCK CHASE

In Sherbrooke Valley, one warm afternoon, I came suddenly upon a man of middle-life, sitting in the shelter of a bush, cooling his feet in the spring-clear stream. He was a stranger to me, but we greeted each other as if some mutual bond gripped us. There was at once joint admiration of the loveliness of the scenery, and we began to exchange impressions about the beauties and the appeal of Cannock Chase.

My new-found friend told me it was only in the last three months that he had known the Chase at first-hand, but he had in that brief period made five half-day rambles through Sherbrooke.

"I feel I have discovered a friend whose company I want to be in all the time, and Cannock Chase is the name of the friend," he said.

The friendship of Cannock Chase then became our only theme, and as we trudged on together an hour later towards highways and streets that led homeward, we felt that the friendly Chase had won both our hearts.

It is the assurance that the friendship of Cannock Chase is shared by men and women in all parts of the world that has led to the appearance of this volume as a successor to

The Best of Cannock Chase of two years ago. From distant parts of the Empire there have been received kindly acknowledgments of pleasure derived from reading of never-to-be-forgotten familiar moorland and woodland places, and learning of historic incidents which are landmarks, and of men who are memorable in the story of the Chase. Home pride has been stirred in those who have kept to their native heath as in those who have gone to other counties and countries, and they have hastened to own the friendship of Cannock Chase.

In the pages of this book something may be learned of men who, in their varied callings, circumstances and favourite pursuits, revealed that they were friends of Cannock Chase. Some were natives, others became like natives of the Chase.

A suggestion made in the closing chapter of *The Best of Cannock Chase* resulted in the founding of the new organisation, "The Friends of Cannock Chase," whose chief aims are to assist in the preservation of the natural beauties, features of local interest and open spaces of the Chase for the benefit of the public, and to stimulate a better appreciation of its many amenities.

As soon as the formation of the Society became known, there was a strong rally to membership, and useful work has been done by The Friends in arousing public opinion against undue slaughter of the deer on Cannock Chase, and in calling attention to the urgent need for the protection of public footpath rights.

The Society did all in its power to set in operation influences that might be the means of saving Beaudesert Hall from demolition, but circumstances beyond its control sealed the fate of the mansion.

In days ahead, other issues may arise in which The Friends of Cannock Chase may be able to prove their love of the entrancing and abiding Chase, and their readiness to support all just proposals and enterprises whose object is to secure a fuller enjoyment of the friendship of Cannock Chase.

JIM AND JOE SEE WONDERS

Jim the Rock-Header was on the Chase having a stroll where the breezes blow, and when I came across him he was kneeling in a gravelly gutter which the rain had made in the hollow between two undulations of hills.

Jim was handling a number of pebbles which had been washed clean by the moorland streamlet, and as I got up to him I said, "Hello! Have you found some gold nuggets?"

"I wish I had, but it's my luck to lay hold of nothing but what I sweat for," was his answer.

Jim turned from his knees on to the ground and began to point out to me the varying features of big stones which he termed bibbles, and smaller ones which he named kidneys. He rambled into a lecturette on veins, fissures, curves and angularities of the stones, and it was obvious that although he had no geological learning he had a respectable theory of possible origins and causes.

Jim went off on his magic carpet into æonic realms and made daring suggestions about the likely adventures of the stones which were lying at his feet.

"But it's when I'm down the pit six hundred feet below that I put my thinking cap on most of all," he said, as he toyed with a stone which had blue and red markings upon it.

"It's darnation hot when I'm right away from everybody in a rock-heading, and all of a dither as I hold the compressed-air drill which pokes its nose into the rock. And upon my oath there's plenty of row and dust in the place, but, do you know, I think many and many time a day about what the world was like before the rock was made and before coal was like what it is nowadays."

Once more Jim set out on a journey into bygone millions of years. He pictured illimitable forests, marvellous upheavals of world contents, mighty oceans, sweeping torrents and irresistible glaciers. When he had worked his imagination for several minutes he picked up a bibble and, crashing it on the kidneys, exclaimed: "And that is how I come back to earth again after my little flight."

"I am never lonely or miserable in the rock-heading," Jim went on to say. "I love to fancy how things used to be afore any of we chaps arrived on earth. You can take it from me there is nothing like having something interesting to think about when you are a couple of miles from the pit-bottom. My old dad, who worked in the same pit as me for nigh on fifty years, always told me that an empty brain-box was a killing sort of thing for a chap to have when he is all on his own in a heading for hours on end, and thee cans't be certain that I agree with what he said.''

Jim gave me a hint that the next time I went down his pit I must be sure to go into his district and see him working with the drill.

A few days later I saw the rock-header in his element. Piles of jagged debris were around him, and he was still bringing more down.

"These are the things I love to come across,'' he said, as he called my attention to the delicate tracery of leaves on a flat piece of rock which he had placed for safety on a ledge of roof-timber.

A few minutes later it was "snap-time,'' and Jim and I sat down on the rock-heap. As he drove a heading with a clasp-knife into his chunk of bread and cheese he resumed the chat we had in the gravelly bed of the Cannock Chase moorland rivulet, and he found illustrations for his inventive thought on the roof and sides about him.

. . . . .

On Hednesford Hills, in sight of the Valley pit stacks, one day I met another miner visionary. His working days were over, and he often loitered on the moorlands that gently rise to the reservoir. He was a stumpy, wizened, curved back veteran who leaned on a rough, knotted stick. His grease-stained bowler hat rested at an acute angle on the back of his head; an attenuated and grimy clay pipe was between his teeth remnants, and his beard was mingled grey and rusty.

He stood with his back against an old boundary post, and with him was a younger man who was being interrogated.

As I approached I heard the query: "But what about the kingdom, my lad! What about the kingdom, do you 'ear?"

As soon as the man with the stick, the pipe and the bowler saw me he thrust his arms up in the air, his hedgerow stake narrowly missing the nose of the catechised companion as the right arm made its dramatic sweep.

"I've seen 'im! I've seen 'im. Bless 'im, I've seen 'im."

"Whom have you seen?" said I.

"Thou knowst who I mean," came the impatient reply. "I've seen Him twice this morning. It's the Master I mean. As I come round the bend in yon path down by the Valley 'e stood close by a goss bush that was just coming into bloom; 'is face was ever so sweet, so smiling, and he said to me, 'Joe, I shall want you to come to me afore long. I know you are getting a bit tired, for the road's a long un.' "

"You are getting on for eighty, I suppose," I broke in upon the old man's monologue.

"Ah, 78 last back-end, but thank the Lord, I can still mek a pound of beef steak look sick when I get a chance."

Having replied thus, Joe reverted to more ethereal things.

"And after I come by the goss bush, I was creeping along when I heeard a lark strike up, and, begad, when I looks up at the bird, all I could see was 'is face agen. Upon my life, it was like angels I heeard singing, and 'e talked to me agen, and I felt I loved 'im more than ever I'd done all the eight-and-fifty 'ears I've know'd 'im."

When the tears began to steal on to his cheeks I knew the climax of the visionary had arrived. The gorse bush, the lark, the angels and the Face had been but accommodating aids towards the accustomed end.

"And now 'ave you got an old pair of trousers you can give me?" was the blunt question which brought us back to common earth again.

"Or a pair of old shoes? An old jacket wouldn't be amiss, per'aps, and if you can leave a copper or two in the pockets, it would be none the worse."

ALONG THE BOUNDS OF ANCIENT CANNOCK FOREST

Wandering one day into a secluded hollow, of whose existence I had not previously known, I found myself in a thicket of riven, malformed decrepit oaks whose years of leaf bearing had been left decades behind.

I felt my imagination stirred, and in fancy I heard these patriarchs of the ancient Cannock Forest lisping broken words of apologies for lingering so long in that sombre solitude.

Taking my bearings, I estimated that hereabouts was the heart of Cannock Forest in the days when it was a royal hunting ground. Playful thoughts could have carried me away into scenes of mediæval kills and thrills, of kingly and knightly warfares for possession or exercise of rights; in every century cycle there could have been found some noted incident of desperado and devotee to crowd my mind as I sat musing on the decayed limb of an oak whose branches swayed and creaked in the winds when King John rode that way.

As a great adventure in reconstruction of dimmest days I might have listened to the sounds of litanies of Druidical priests as they made their weird offerings in some such forest grove. My thoughts, however, swept far and wide from that remote retreat to the streams, the bridges, the tracks which in ancient times marked the bounds of Cannock Forest.

I roved from the Tame in the East to the Penk in the West, and from the Trent and Sow in the North to curving lines in the South which touched Coven Bridge, went through the centre of Wolverhampton, then made acquaintance with Bilston, Wednesbury and Bescot, and took a firm grip of Walsall.

Aldridge was brought into the chain of places; a mile or two south of that town Bourne Bridge was reached, and so on the road between Sutton Coldfield and Drayton the bounds touched the Tame at Tamworth.

By Tame to Trent, Wichnor came into view, and King's Bromley, Rugeley and a dozen other landmarks sheltered themselves within the Forest bounds. Stafford, with its walls, held itself aloof a mile or two away.

Such was the extent of Cannock Forest when Courts of Foresters and Verderers with Juries of Twelve Knights in the thirteenth and fourteenth centuries based their affirmations of the Forest limits on the observations of perambulations and confirmatory evidence placed before them.

There were changes in bounds from time to time, and protests arose from the people when the Royal Forest was extended south to the Watling Street, and the resultant application of the Forest laws became irksome.

Royal Forester, Forestership of Cannock, Capital Forester, Censorship of Cannock Forest, Stewardship of Cannock Forest were some of the terms used to describe the offices held by those invested with authority. The clashes of rivalries between hereditary claims and ambitious plots have filled many pages of local chronicles over several centuries.

Cannock Forest had to compete with Cannock Chase in nomenclature when, at the end of the twelfth century, the Bishop of Coventry and Lichfield's Chace, or hunting ground, came into being within the bounds of the Royal Forest. The Bishop's Chace included the manors of Cannock and Rugeley, and was granted by Royal favour as a free Chace for the Bishops and their cathedral. The Chace became free from the severities of the Forest laws, and life for the people in its area gained in circumstance and prospect.

Quarrels took place between Kings and Bishops about exercise of privileges in the Chace portion of the Forest, and the records reveal many disputes about encroachments beyond defined boundaries.

Cannock Forest had nine hayes, or divisions, and when the hayes of Cannock and Rugeley had been handed over to the

Bishops of Lichfield seven forest hayes remained. In the time of Henry the Eighth, Leland, the noted antiquary and topographer, called Cannock Forest the Forest of the Seven Hayes. Teddesley Hay, Cheslyn Hay, Ogley Hay, Gailey Hay, Bentley Hay are local place-names which link the present with the past story of Cannock Chase.

Other Forest names and landmarks also remain, and old families have continued their associations with the district down the centuries.

Great developments have taken place in the Cannock Chase area in the last two centuries, and enclosures for various purposes have considerably reduced the open Chase. In the early years of the nineteenth century the Chase was estimated to comprise about 32,000 acres, of which 18,000 acres were enclosed and cultivated, and 14,000 acres were open waste. Year by year builders and cultivators take away slices of the breezy moors, and Forestry men cover up vast expanses with their saplings.

Cannock Chase has arrived at a critical stage in its history. A watch must be kept upon those whose projects and claims are likely further to diminish the beauties, the amenities, and the free access to the glorious Chase.

SILVER BIRCHES AND A LOVELY SKY

"What a lovely morning!" was the remark made to a miner who was taking his dog for a stroll.

"A lovely morning, did you say? Why, I should call it more than lovely; it's grand, to be sure," was his answer. "Give me this rather than scrawling that couple of miles to my stall underground that I have to do every morning," and then he proceeded to describe where he had been for his constitutional. He had been out to Brocton Gate, all over Anson's Bank, right across Penkridge Bank, and was now making for Prairie Farm. Twelve to fifteen miles on shanks' pony put him in form for his Sunday dinner.

"How pretty the silver birches look today," he added; and then he asked if I had noticed what a pretty sky it was that morning.

"I say, do you ever see all sorts of things when you look at the sky when the sun is rising or when it is setting?" was his next conversational diversion. "I often think I can see mountains and lakes up there, and as the clouds move about they make me think of places I went to in the war. My eye, it's real funny to watch the shapes and the colours altering all the time. As soon as you have made out one bit of country, it gets wiped out and something else makes itself, like them comic pictures on the screen. Sometimes when me and the dog are out on our own I get so taken up with the sky that I forget all about Jimmy. Now, tell me, I suppose you get a bit odd at times about the sky, like me?"

Having made my "Of course" admission, we reverted to the silver birches theme, and agreed that on that sunny, frosty February morning the trees had a rare loveliness.

Right along the roadside for some distance was a succession of graceful, feathery-topped birches. An infinite variety characterised their sinuous outlines. Even in their silvery barks there was a delicate diversity of tone. Around their boles were rows of stippled hieroglyphics that seemed like a Creed message of faery import.

Against the sky background some of the trees looked like the thousand-fold tributaries of some huge far-expanding Mississippi—main sweeps of torrent, lesser rivers and remote rivulets, traversing the vast spaces until the ocean rush was reached.

Devoid of green, except the mossy coat which wrapped itself about the roots, the birches yet possessed a beauty and a symmetry that could not be matched by any other trees that neighboured them, and the slender catkins carried a film of frost that added to the decorativeness of the silver lustre-bearers. Ranked on the edges of Pottal Pool were hundreds of birches, whose height of from thirty to forty feet gave them a stateliness that was not attained by the road-side Indian-filers, and, unmoved by any breeze, undisturbed by a solitary bird, unaccompanied by the ripple of water, their mood was one of supreme serenity.

At their back rose in ascending levels lines and lines of larches and firs. The rotten mush of last year's bracken was littering the ground. Pale green mosses made damp patches beneath the decayed brown. Desolate clumps of gorse were still carrying the rags of past blooms. On shabby remnants of heather could be detected faint evidence of petals which once were purple.

Four miners came along with their dogs. As they passed by they had a word of admiration for the silver birches. They, also, said it was a grand morning, and one of them even added a comment or two of how much nicer the Chase would be looking in a few weeks' time.

BOGS TURNED INTO FERTILE LANDS

Staffordshire has no greater romance of its kind to offer than that of the wonderful transformation of Cannock Chase bogs into fertile lands by Edward John Littleton, of Teddesley Hall.

When he succeeded to his uncle's estates in 1812, he set to work to make them the best in the county. He took upon himself the task of reclaiming several thousand acres of wild Cannock Chase lands, which was only possible of achievement by the spending of a large sum of money on drainage.

Littleton counted the cost and tackled the work with skill and courageous enterprise. He declared that he knew no other pleasure so satisfactory as rendering land more productive and providing a source of employment for years to come. Said he: "My joy in this work is like that which saints must feel in the moral reclamation of a vicious character. I must, however, own that in this case there is the anticipation of double mammon. I not only make the land fruitful, but make the devil I have exorcised do work of a slave at my mill wheel. The water thus obtained will alone repay the cost of acquiring it."

Littleton drained his reclaimed lands to a depth of twelve to fourteen feet. By 1844 he estimated he had added three thousand six hundred acres to his estates at a cost of £120,000, and he wondered what old Sir Edward Littleton, his uncle, would say if he were to rise from the grave and see flourishing lands where previously there were what were thought to be irreclaimable bogs.

It was the first time these lands had ever been broken up, and landowners throughout the county and in other parts of the kingdom visited the Chase to see the marvels Littleton had worked. Men of science, agricultural experts, statesmen, all made tours of inspection and studied data concerning the methods of cultivation being adopted. In the pages of his MS. Journal, from which I have been able, by the permission of Lord and Lady Hatherton, to quote many passages in this book, are many references to these visitors.

Littleton was a pioneer in his scheme for the employment of boys on the land, and his insistence on their production of a certificate that they could read and write before they were put on the land ensured that they were fit to commence in an industry in which brains would be required. New implements were invented at Teddesley which aroused great interest when exhibited at Royal shows. Littleton went to London and bought old post horses for ploughing, and the animals became the admiration of all who saw them.

His cattle gained a remarkable repute, and the Teddesley sales attracted buyers from all over the kingdom.

"My rickyard is a glorious sight, and not yet what it will be; I have carried wheat with fifteen carts or waggons all day," said he one Autumn, and in the memorable year of 1843 he devoted much space in his Journal to a graphic picture of the crying of "Harvest Home." It was after the shortest, most favourable and least expensive harvest he had ever known. He believed his rickyard to be nearly the largest in England, probably the largest. All his crops, except one field of barley, had been gathered in without a drop of rain.

He gathered his labourers in front of the house, and regaled them with a full quart of ale each. First they drank "Lord and Lady Hatherton." Then they assembled in a large ring, the farm bailiff in the middle, and gave three shrill cries, allowing time between so that the echo could return from the Beech Wood and The Hayes.

Then the men gave three tremendous cheers, then shrill cries again, and again cheers. This was followed by the drinking of the health of "Mr. and Lady Margaret Littleton," with the same cheering. Then Old Bellamy, the oldest waggoner and ploughman, was first in the middle of the ring, and instead of shrill cries he repeated between each three cheers the words:

"Well blowed, well sowed,
Well reaped, well mowed,
And ne'er a load o'erthrowed."

After the cheering was over they each time resumed their seats on the wall. Then the toast was "Our noble selves," followed by "Queen and England," and more cheering, with the verse given above repeated between each cheer. Then they had a fifth horn each and went home.

Littleton adds the explanation that harvest home was cried when the last waggon left the field. "I heard my neighbours crying it a full mile or two off the other night. I dare say the cry to-night was heard at a much greater distance." It must have been heard at Penkridge, and probably the people of Cank listened to the far-off sound.

TALKING OF PLATO BENEATH A COPPER BEECH

"Come and see my garden and my bees, and then you can dip into the old Minute books of the Alport Trust," was the message I received one July day.

It came from Bednall, where the Rev. A. R. Alsop, M.A., had been Vicar for over forty years. I had often seen him drive into Cannock in his governess cart to attend the meetings of the Cannock Board of Guardians, of which he was vice-chairman for sixteen years, a position which gave him much pride, for he served under the distinguished chairmanship of the third Lord Hatherton.

He often brought with him a roomy basket covered with a nice white cloth, and I soon learned that it was not his lunch that he brought with him, but kindly gifts for the sick folk in the Poor Law Infirmary.

When I walked over the Chase to spend a day with him as arranged, the Vicar of Acton and Bednall met me at the cross-roads at Bednall Head. We went past the low-roofed cottages in whose front gardens summer flowers were all aglow; opened gates --and faultlessly closed them behind us—as we walked out of main lane into bye-lane which leads to the Vicarage, hidden away amid bushy hedgerows and tall trees.

First we enjoyed cool shelter in spacious rooms, where family heirlooms were prominent as adornments. There was a picture of a quaint, mediæval homestead at Alsop-en-le-Dale, where the Vicar's forefathers had their origins, and we made surmises regarding the variant spellings of the family name, with their whimsical preferences for one or two l's, s's and p's. In the hall was a painting of the earliest Bednall Church, with its unattractive elevation and meagreness of architectural features.

A leisurely hour was spent in the garden, where the pomp of summer was resplendent in borders of many colours. A row of hives transformed one corner into a zone of incessant buzzing, and the Vicar confessed he sometimes likened it to an "industrial hereditament"—a phrase that figured in his Assessment Committee deliberations.

As we sat under the shade of a copper beech, the wide expanse of Teddesley Park was spread out before us, and memories were awakened of the days when Mr. Alsop's father came to that obscure spot as parish priest in 1867. Thirteen years later son succeeded father, each of them complying with the condition that the living must be held by an Exhibitioner of Brasenose College, Oxford, and for three years the son learned under his father's care the gentle arts of pastoral oversight by acting as curate of Acton. For sixty-one years did the faithful rule of the two Alsops continue.

"It has been a quiet corner of Arcady, but a lovely one for me," said Mr. Alsop. "My bees and my flowers, my work of coaching students, my duties as a Poor Law Guardian and Rural Councillor, have kept me busy, and I would not have changed my lot for that of any parish priest in the kingdom. In some little way I have tried to bring about some of the ideals enunciated in Plato's *Republic*, which was my friend day and night when I took my degree in political economy at Oxford."

It was not to be expected that antiquarian matters should remain outside our conversation. We discussed local place-names: Acton—Oak Town, of which there were thirteen in England—and the Anglo-Saxon Bednall.

"Just imagine for a minute with me," said he, and then the Vicar described the humble dwelling of the chief named Beda who ruled thereabouts before the days of the Conqueror and Doomsday Book. A vivid picture was given of the rugged-countenanced chief with his fair hair and blue eyes, going out into Cannock Forest with his tall sons and shaggy dogs to slay the wolf, or to bring back a fat buck from Berkswic. A hostile horde overwhelmed chief and sons, and all that remained of him to-day was the name of Beda's Hall or Bednall.

Later in the day we went to see Acton Church, with the date 1566 over its west window, and the ancient Moat House with its many associations of the Trussells. A visit was also made to Bednall Church, standing on the same hill where the former chapelry was situated. As we walked about the lanes we talked of the comparative excellencies of ancient and modern stained glass windows, the skill of "Good Mr. Street," the restorer of Acton Church, and also of the Law Courts in London; the meaning of the word "waste" when used in Cannock Forest records, the discovery of an old font being used as a horse-trough in the parish, and the presence in the vestry at Acton of a bassoon, which bore testimony to days when minstrels led the singing.

I listened intently as the Vicar spoke with real gladness of the production of the choicest butter and honey, and laughingly told stories of his experiences when judging exhibits at County Shows. He related reminiscences of his public life, and now and again he would strike impulsively into far realms of thought as he gave his conclusions on the influences which went to the making of the glory that was Greece and the grandeur that was Rome.

While the Vicar did odd jobs about the garden in the evening, I spent an hour alone in the library delving into the Alport Trust records. For two hundred years there had been vicars on and around Cannock Chase who had drawn good sums of money from the Trust, and here I found their signatures for benefactions received each St. Barnabas Day. One of the unexpected discoveries was that for thirty-four successive years two former vicars of Acton and Bednall, father and son, received the Alport money, even though they had failed to fulfil the Trust condition of residing and working in the parish.

This revelation provided us with an ample topic of conversation for the last hour of the visit, and when I prepared to leave at the end of a most enjoyable day, the Vicar insisted on my acceptance of a bundle of volumes of the Staffordshire Parish Register Society. Into my pocket he slipped a choice leather-bound copy of John Keble's *Christian Year*.

It was a full-of-smiles farewell that he gave to me, and his countenance had that apple rosiness which shed radiance wherever he went:

> That comely face, that clustered brow,
> That cordial hand, that bearing free,
> That exquisite serenity.

MR. ALBERT STANLEY, M.P.

STANLEY'S STORY OF THE MISSING CANDLE

In the century that has gone by since the valleys of Cannock Chase began to be peopled by colonies of miners no man has won his way to the hearts of his fellows as did "Albert," as he was familiarly called.

The son of a miner who lost his life by an explosion at a Hednesford colliery, Albert Stanley also himself had experiences in the pit which gave him abiding impressions of the dangers of coal getting. At nineteen years of age he received severe injuries which included crushed ribs, a lacerated skull and a smashed ankle, and for the greater part of six months doctors despaired of his life.

That was the end of his career in the pits, and when, several years later, the miners of the coalfield required a champion of their cause for better hours and pay they chose "Young Albert," of twenty-four, who had spent his leisure hours to a profitable purpose, and whose fluency of speech and skill in marshalling figures and stating a case had astonished everybody.

The fact that Stanley, like so many of the miners of the Chase, had come a few years previously with his father from Shropshire—Dark Lane was his native place—added to his popularity.

The moment it was known that "Albert" was to speak a crowd of miners rushed to hear him. It was so at the beginning of his leadership, and it was the same twenty-five years later. He had a marvellous way of swaying a mass meeting of ten thousand, just as he could mould to his desired point of view negotiators at round-table conferences. Sometimes critical days developed, the men became restive because of overtures which were prolonged and apparently fruitless, there were threats from some of them that when the mass meeting was held in the twenty-five acre field at Heath Hayes they would tell "Albert" what they thought about him for allowing things to drift for so long.

On one such occasion I had arranged to meet Stanley and walk from Hednesford to Heath Hayes with him. He had been attending to his Parliamentary work in London for some days, he having been elected member several years previously for North West Stafford division, and wished to get a few impressions of the tone of things locally from one who had been moving to and fro over the coalfield during the past week.

He understood some of the men were inclined to make it awkward for him, and that there might be some disturbance at the meeting.

"But you wait and see. You will find I shall have the last say. I have been through this sort of thing too often not to know how to handle them. When these chaps try their game I shall just twist them round my little finger like this,'' said Stanley, and he suited the action to his words.

The crowd was a huge and expectant one. Before Stanley began there was an ominous buzz, and dozens of arguing knots of men. As soon as the miners' agent stood up there was complete silence, and his first words were so good-humoured and confiding that the grousers were disarmed. For over an hour Stanley continued his winning way, and not a solitary disgruntled miner impeded his progress.

Stanley took hold of some of the objections that had come to his ears, and turned them inside-out until they became in his hands nothing more than natural difficulties of interpretation which could be explained with full satisfaction to all concerned.

After a quick survey of the situation and the clinching influence of an eloquent peroration a resolution of Stanley's drafting was unanimously passed without a dissentient voice or hand. As the miners dispersed from Five Ways and went their fifty roads home there could be heard on all hands the remark: "Well, what do you think about it? Albert on top again; he beats the grumblers every time.''

"What did I tell you, my lad? It all happened just as I knew it would. It was fine to stand up there and see them nodding their heads with approval just as if I was pulling them all at the same time," was Stanley's greeting to me as he came down from the waggon.

"Tell them the story about the lad and the candle," said Sam Cotterell one night to Stanley, at Old Hill, when a meeting was being held to appeal for help for the Cannock Chase miners who were out on strike.

It was an incident the news of which had come to hand the previous day. A teacher at a Hednesford elementary school had missed a candle out of the classroom. He tried to find out who was the culprit, and at last a little lad put up his hand and said, "If you please teacher, I had it and eat it because I was hungry."

Stanley narrated the story in the telling manner in which he was so accomplished, and by the time he had explored the pathos of the incident there was not a dry eye in the building.

At the age of fourteen, Stanley passed his examination as a local-preacher, and was known as "The Boy Preacher." That was the earliest training he had to equip him to be a miners' leader, and in the Methodist chapels he unfolded his favourite blend of the Sermon on the Mount and social reform. There was hardly a chapel in Salop or Staffordshire that did not get Stanley to preach for them, and over a period of twenty-five years he did not have more than five or six free Sundays.

In the strike periods he would address public meetings in the morning and afternoon, and then preach at night. Large sums of money were raised for the miners and their families, and so heavy were the boxes and bags that the collectors had gruelling journeys back to the Chase with their loads. It would have been wicked to have changed the pence into more compact forms for convenience of cartage, for that would have destroyed the imaginative appeal which the sight of the piles of coins would make on those who were waiting to count them at Hednesford.

One winter's day Stanley took me to several out-of-the-way corners of the North-West Stafford constituency where he was fighting to retain his seat. There were little National schools, with meagre oil-lamps, and uncomfortable desks into which the country folk crowded, and I was able to see the miners' M.P. exercise the spell of his speech as felicitously as I had on other occasions in vast open-air gatherings, and in packed halls and chapels.

Many were the demonstrations I had of Stanley's skill in surveying events and discussions in a lucid précis which could be transferred straight to the printed page without the least alteration. Nobody need ever be in doubt of the meaning of anything he said. Life is too short to say a thing badly and then have to explain what you mean by it, was one of his sayings.

In Parliament, County Council, on the platform, in the pulpit, in the open-air, or in the conference room he had a clarity and sure touch; aims founded on principles enthralled him.

His love of books began when, as a lad, he went to Oakengates, and out of his first earnings in the pit bought Morrell's Grammar, Paley's Natural Theology, and a shilling edition of Shakespeare. When he became miners' agent books were his acknowledged extravagance and his library grew from three to three thousand volumes. All the great Victorian novelists were on his shelves; Ruskin was there in forty-five handsome volumes of a definitive edition, and first editions of "Pamela" and "Oliver Twist" were among his treasures.

Stanley died at fifty-two. There were drawn blinds in every mining village of Cannock Chase; veterans of the pits and their sons too felt they had lost a great champion.

I take down from my shelves a book in which Stanley's name is inscribed. It contains a copy of a letter which Stanley wrote to the widow of Dr. Charles A. Berry, of Wolverhampton, who worked magnificently in the 1893 strike in raising funds for the miners of the Chase.

Stanley's pencil mark is by the paragraph in which, on behalf of the miners, he pays a tribute to Dr. Berry; "He had a tender heart for the struggles, temptations and sorrows of the people; all the great wealth of his intellect and wonderful witchery of words were again and again employed against the inequality and cruelty of our present economic and industrial system. He did not flatter the people, for he knew their weaknesses and faults too well, but his sternest admonition was always tempered with the spirit of Christian sympathy and love that won for him their warmest admiration and regard."

That is in the typically Stanley style and tone, and what he said of Berry aptly sums up the character of the man who wrote it.

CANNOCK CHASE DEER FOR A THOUSAND YEARS

Cannock Chase oaks and Cannock Chase deer have for hundreds of years been famous in local annals, and a great deal of the history of the district gathers around them. They appear in the chronicles of the Chase side by side—the majestic and the graceful—contributing to the romance, the tragedy, the hard struggles, the joyous sport.

A thousand years ago the hindebrook and the harts' wallowing place were well-known to huntsmen in Cannock Forest, and deer leaps, deer parks and deer slades became familiar features.

Visiting Wolseley Hall one day, I had the privilege of inspecting an ancient document which takes us at once back into the fifteenth century, when Ralph Wolseley, who was made a Baron of the Exchequer in September, 1468, was given power to enclose a park and stock it with deer from Cannock Chase. It was Edward IV. who made the grant, in the ninth year of his reign, Wolseley's privileges including the freedom to construct deer leaps, which, while allowing the animals to get into his park, prevented them from returning to the forest.

This Royal document hangs in the entrance of the Wolseley mansion, and it still bears the Great Seal.

Where deer were concerned extraordinary incidents were always likely to occur, and it was no uncommon thing for the estimated law-enforcer to find himself charged with a forest offence. Hugh de Loges, Chief Forester of Cannock Forest, was imprisoned in London for taking venison from the Forest, and the King decided it would be better for all concerned if he took over the control of the Forest.

The stag or hart was regarded as the real beast of the Forest, and until the reign of Henry II. the penalty for killing one was the loss of life or members. Gradually the punishment became less ferocious. King John was a stern punisher of those who stole his venison, and he had no hesitation in exacting dire penalties from deer stealers.

It was a crime to keep unauthorised dogs and weapons, to construct deer leaps, or make illegal enclosures. Mastiffs kept by farmers and others had to submit, every three years, to having the claws of their fore-feet cut off with a chisel and mallet, so that they would be incapable of chasing the Forest beasts.

Men of high position were proceeded against for violation of the Forest laws, parson and knight being equally guilty of stealing Royal deer. It was complained in 1264 that three bucks and a doe were taken from the King's Forest of Cannock, and the venison removed to Chillington, the home of the Giffards. A deer-stealer, who was caught in the act, climbed into an oak tree and defended himself by shooting arrows at his pursuers. Two foresters, who caught a man in possession of a roebuck, accepted three shillings not to report the matter, and then took the venison home for their festive board.

A dozen men, bent on having a sumptuous feast for Christmas, took ten does, three bucks and two hinds in Cannock Forest. When the arm of the law began to reach after them, two of the men fled to the Holy Land, and their choice of such a place of refuge was accounted unto them as creditable pilgrimage and they were given a free pardon. Two of their fellow-culprits, who went for safety to Ireland, gained no such excusal of penalty. The men who could not be found were outlawed, two were committed to prison, and another, who was a very poor man and a minstrel, who went from place to place putting sweet music into people's lives, was let off with a caution.

At Staffordshire Sessions, at the end of the sixteenth century, a Rugeley clerk, who had broken into the free Chase of the Queen, called Cankwood, and killed and carried off a deer, had to stand his trial along with a Hamstall Ridware labourer who had broken into the close and park of the Queen, called Beaudesert Park, and killed and carried away six deer on two specific dates, and with Edward Wolseley, gentleman, who was alleged to have stolen one of the Queen's deer from Cankwood.

Witnesses before Court Commissioners in 1530 stated there were always plenty of red, fallow and roe deer on Cannock Chase. John Leland, who came through Staffordshire within the next twelve years, found Cannock Forest to be "a great thinge, and at Shuckborough Park in the side of Cank Woode saw many red deer."

Roger Meuland, Bishop of Lichfield, claimed, in the thirteenth century, to be free to take venison at his pleasure at all times of the year in a circle of fifteen leagues in circuit in the middle of Cannock Forest, and within easy reach of his residence at Shugborough.

Meuland granted Roger de Aston, of Tixall, the title and office of hereditary Master and Ruler of the Game of Cankwood. This office remained in the hands of the Astons even when the Chase passed from the Bishops to the Pagets, and in 1712 the fourth Lord Aston was prepared to relinquish the office on the Recommendation of the Court of Chancery only on condition that every year four fee bucks from the Chase should be delivered to him whole and entire at his mansion at Tixall, without fee or reward, by the Keeper of Cankwood.

In November, 1827, the Marquis of Anglesey appeared at Staffordshire Assizes to give evidence on an appeal against a conviction for deer stealing, and to prove that Cannock Chase deer belonged to him, and that he had given no leave to kill the deer but to persons employed to do so.

In the following November, a Great Haywood labourer was committed to the House of Correction for one month for having attempted to kill, on the unenclosed part of Cannock Chase, a deer, the property of the Marquis of Anglesey, without his knowledge and consent.

In our own day the law continues to exercise vigilance on matters affecting Cannock Chase deer, a Hednesford miner being fined two pounds, at Rugeley Police Court, in June, 1935, for unlawfully snaring a deer near a pool at Brindley Heath. This was the first prosecution of its kind in the district for many years.

WHEN PASTOR JEROME CAME FROM APPLEDORE IN DEVON

When the Rev. Jerome Clapp Jerome migrated from Appledore, in Devon, to Staffordshire in 1855, it was chiefly because he felt the glamour of the coal and iron of Cannock Chase and the adjoining industrial districts. He had heard of the beauty of the thousands of acres of heatherland, and alluring information had come to him of the riches that were to be made from the seams beneath the wild lands in the middle of the county. He hoped to blend his pulpit work with his prospecting in minerals, and while he was active in promoting spiritual progress in Walsall Congregational circles, he had business concerns with the Birchills Iron-works. It was coal, however, that gripped his attention most.

Brownhills and Norton Canes—otherwise Norton-under-Cannock—was a united neighbourhood in the great developments which were taking place, and the men who were hoping to turn the Conduit Colliery into a gold mine were glad to receive the investments of the energetic Independent pastor. Eloquent a preacher though he had proved himself to be, he had not found the calling to be plentiful in remuneration, but his wife,

Marguerite, had inherited a handsome fortune from her father, a Swansea solicitor named Jones. In what more promising venture could her money be placed than in assisting to produce the coal which was to be the life-blood of the industrialism of Mid-Victorian England?

For three hundred years extensive areas of land in the Norton Canes district had been in the possession of the Lichfield Conduit Trust, the income derived therefrom being used for the repair of the Trust conduits and the maintenance of the feeder water-courses in and around the city. Thus it was that the first ambitious mining enterprise at Norton Canes was called the Conduit Colliery. The Bottom Cathedral pit and the Top Cathedral pit were names which also had reference to the Lichfield Trust associations.

Great hopes were held that success would reward the sinkers at the Conduit, and it was an anxious Marguerite who waited for the news of progress brought by her husband when he returned home from Norton day by day.

Pastor Jerome had a courageous enthusiasm, and he kept his preaching appointments on Sundays in spite of his tired and worried body and mind. He had capable and experienced men to help him, but difficulties were constantly arising, and he realised that the capital at his disposal was melting away before the turn of fortune had been reached. Water and running sand had swept more than one man of Cannock Chase to ruin, and Jerome found himself engulfed in such a tragedy.

Late one night, when all at home had gone to bed, the pastor-prospector returned from Norton Canes bearing the bad tidings, the fear of which had haunted his wife like a spectre for weeks. He sat himself on the edge of the bed and broke to Marguerite, as gently as he could, the news that the pit was flooded, and that he was a ruined man.

May 2nd, 1860, was the day of disaster. Some days later Pastor Jerome, leaving the family fortune lost in the rush of waters and his own surname written in large letters on the pit for future generations to read, went to London in the hope that he would be able to get on his feet again. In due course, he was joined in London by his wife and children, including the four-year-old son, Jerome K. Jerome, who was destined in subsequent years to win considerable fame as the author of *Three Men in a Boat* and *The Passing of the Third Floor Back.*

"Papa's railway," as it was termed in the Jerome family circle, was a later venture of the ex-colliery owner. It led "from Poverty to the Land of Heart's Desire," said Jerome Jerome the Second. It did not pass beyond the stage of phantasy, and when bad news was once more told to the ever-hopeful wife, she wrote in her diary : "Every effort my dear husband makes proves unsuccessful. We seem shut out from the Blessing of God."

The name "Jerome's Pit" has stuck to the Conduit as the most frequently used of the two. The miners have never tried to pronounce the name in the three-syllable form which the pastor from Appledore declared to be the correct one.

"He used to call himself something like 'Jerrymy,' " said an old miner whom I met one day in Norton. This veteran was born in 1846, and he described to me how he saw Jerome many times on the pit-bank, where there was a lot of bustle and noise when the sinking was going on.

"A nice gentleman, but he was quiet and used no bad language," was the old miner's single sentence summing-up of Jerome.

Men with deeper pockets than Jerome took over the Norton Canes sinkings. They triumphed where their predecessor came to grief, and the Holcroft family were able to establish Jerome's pits in the forefront of the undertakings of the Cannock Chase coalfield.

MACAULAY'S DAY OF PRAISE ON CANNOCK CHASE

It was on a lovely day in August that I set out from Teddesley Coppice in an effort to walk in the same track that Thomas Babington Macaulay, famous poet, essayist and historian, took over Cannock Chase ninety years before. I had learned from the Hatherton MS. Journal that on that occasion Macaulay spoke for a whole hour in praise of the beauties around him.

I recalled that on the day when Macaulay set out with Baron Hatherton for their Chase tramp to Beaudesert, they passed through the farmyard which was packed to overcrowding with crops from the wide plains which had been reclaimed from the wild land of the moors within the previous thirty years.

They saw massive shattered oaks in the woods between the hall and Pottal, which were survivals from Tudor times. Beeches, firs, pines, which had been planted for the dual purpose of beauty and profitable investment, rose out of the dense undergrowth of briars and bracken.

Deep ruts made progress for them slow through the plantations, and a rush of water could be heard as it dropped down the weir at Pottal to hurry away to turn the mills at sundry farms between Ivy Farm and Otherton.

Deep fissures had to be nēgotiated in Pottal trackway, where, a few days previously, the rains had hurtled from the hills towards the lake. On either side of the rough track were unkempt gorse bushes and slender birches, and on the slopes towards Pye Green there was a wide expanse of open lands. Reaching the top of the ridge at Boss Green crossing, Macaulay had his first glimpse of the hills at Wandon, where Beaudesert Park could be entered.

It was then that Macaulay burst into poetry. They were lines of the sixth Canto of Scott's *Lady of the Lake* that he recited :

> " 'Little we rack,' said John of Brent,
> 'We Southern men, of long descent;
> Nor wot we how a name—a word—
> Makes clansmen vassals to a lord :
> Yet kind my noble landlord's part—
> God bless the house of Beau Desert !
> And but I loved to drive the deer,
> More than to guide the labouring steer,
> I had not dwelt an outcast here.' "

Macaulay and Hatherton came to the conclusion that Scott must have made the acquaintance of the "Waterloo" Marquis while the latter was staying in Scotland, and that the introduction into *The Lady of the Lake* of a reference to the Paget family and place was a courtesy of the poet to his new and very striking acquaintance.

As the two friends walked together over the moors in the direction of Wandon Lodge, Macaulay continued to draw from his vast store numerous poetical passages applicable to the scenery which was around them, and expressing sentiments which contemplation of it provoked.

"What a wonderful memory," remarked Hatherton, "and how singular that you should be able to quote those lines from Scott as soon as you caught sight of the Paget lands in the distance."

Macaulay assured his friend he had never read the lines in question since he was a boy of thirteen at school, thirty-six years before. It was quite true he had a good memory, for he was certain he knew by heart forty thousand lines of English poetry, and could repeat by heart more than half of Milton's works. If his two volumes of the History of England were by some mishap destroyed, he could write them out again without a note to help him.

Stile Cop and Castle Ring stood out in front of Macaulay and Hatherton as they pursued their course along the Marquis's Drive, over the Rugeley and Hednesford road at Moor's Gorse. Inside the park were magnificent hawthorns, and Hatherton was provided with a literary feast as his guest again quoted favourite passages from the poets.

Macaulay did not find Beaudesert to be the abode of the Muses, and there was little affinity between Anglesey and the poet-historian. The old warrior, however, paid marked attention to his new visitor, whom he showed all over the mansion, and gave him and his neighbour from Teddesley a handsome luncheon. Anglesey was particularly careful to show Macaulay his picture of the Cavalry at Waterloo, and to tell him it was a personal gift to him from George IV.

Pleased with his visit, and enjoying his renewal of open-air breezes, Macaulay resumed his literary rhapsodies about the beauties around him. On the way back to Teddesley he maintained a stream of recitations and stories. In these he displayed a marvellously intimate acqaintance of many scores of passages in which poets had praised the landscape glories of England, and with notable incidents and personalities in Staffordshire story.

That same night, after dinner, Macaulay captivated everybody at Teddesley Hall by reciting many old ballads which had been written about notorious crimes. For hour after hour he went on, and the news of the blood-curdling tales he was telling soon spread through the house. The servants came from the kitchen to listen at the doors, and Macaulay immediately called to them : "Come inside and you will be able to hear better."

Inspired by the rapt attention of such an audience, Macaulay revelled in the dramatic light and shade, and so thrilled were his listeners that there were many heated heads on the pillows in the maids' room that Autumn night.

THE BLUE PUMP AND THE CATHEDRAL

"I could have eaten my head off when I heard they had pulled down the old Blue Pump, and I hadn't had a chance of looking into the cellar," said a miner to me one day. The reason for his regret was that his grandmother, who had lived there at one time, had told him of the underground passage that went from the old public-house to Castle Ring, and all the way to Lichfield cathedral.

His grandmother had also told him of older men who had been part of the way into the tunnel, and they had even said Dick Turpin used to call there.

In reply to a question, he told me it was not his grandmother but another occupant of the Blue Pump who could boast that in the course of his married life he had lived in no fewer than forty-five houses.

I thought of the Turpin and cathedral legends when I went to explore the curious features of the Blue Pump. The building had fallen into a dilapidated state, and it had been doomed to demolition by the local council. It had long before ceased to be a licensed house, and no one living seemed able to state when last the inn catered for travellers between Cannock and Rugeley, and harboured men of the cock-fighting, prize-fighting and horse-racing fraternities who frequently met on the adjoining Hednesford hills.

The miner who was about to be dispossessed of hearth and roof soon informed me of the tunnels and Dick Turpin. Old Dick Benton, the Hednesford blacksmith, who lived to be eighty-six, told the miner's grandfather all about them, and Dick heard it from his grandfather. So there was no fear of it being little more than an idle tale.

We included the cellars in the first part of our tour. By candlelight we wandered about, poking into every nook and corner. Litter was plentiful, but except for structural features no signs of tunnels could be detected. There was a small curving of brickwork on one of the walls, and when I suggested that this might possibly be where the drainage of the cellar made its way into the sub-soil or by a gulley into the stream from Hednesford Pool, the miner ridiculed the idea, and pinned his faith to the cathedral tradition.

As he had been a tunneller in Flanders during the war, I asked the miner if his experience had taught him that it would be impossible for anybody to journey underground from Hednesford to Lichfield for seven miles in an unventilated tunnel. He answered by telling me of another tunnel which went from Castle Ring to the cathedral at Lichfield. He had also heard of one that went from Leacroft Hall to Cannock Church, and of another that connected Cannock Church with Penkridge Church.

It was useless for me to try to rid his mind of the local tunnels beliefs, but he was not so obstinate, after listening to historical evidence, about the Dick Turpin tale. He was prepared to dismiss from his mind the statements that several horse-shoes which were nailed to one of the stout beams in the Blue Pump kitchen were made by a local blacksmith for one of the highwayman's steeds.

"But I shall still stick to there having been a tunnel to the cathedral," he added.

HATHERTON HALL.

LITTLETON'S KEEN EYE FOR BEAUTY

Mary Hyacinthe Wellesley, niece of the Duke of Wellington, who married twenty-one-year old Edward John Littleton, of Teddesley Hall, was one of the society beauties of the day, and it was not surprising that the rich young Member of Parliament for Staffordshire should at once request Sir Thomas Lawrence to paint her portrait. That picture may still be seen adorning the walls of the Littletons at Hatherton Hall, Cannock.

In the last two hundred years of Cannock Chase history, no names stand higher for influence wielded for good on the life of the district than those of the distinguished pair who became the first Baron and Lady Hatherton.

Theirs was an ideal love match, and in the many hundreds of pages of Littleton's MS. Journal are numerous entries in which acknowledgment is made of the great part which the wife played in the moulding of her husband's purposes, and of the undiminished affection maintained between them over a long period of years. The Journal contains love annals which never lose their freshness and their charm from the first page to the last.

In the Autumn of 1812 everybody in Brighton was talking about the presence in their midst of the five reigning beauties of London. One of them was Hyacinthe Littleton. Littleton saw all the five thirty-five years later, and said: "The brightest is certainly greatly faded, but there is more freshness left about my flower than the rest, and much grace about all of them."

Emperor Alexander I. of Russia had no hesitation in declaring Hyacinthe Littleton to be the handsomest woman in England.

Attending a ball in Naples many years later, Littleton came to the conclusion that there was only one handsome lady in the room with the exception of Lady Hatherton.

Hyacinthe was always the handsomest woman in the world to Littleton, but he had an eye for the beauty of other women also. She, on her part, was flattered to feel that she held the field in all the comparisons her husband made.

Littleton had a memory revived at Wolverhampton railway station in July, 1843. He there saw Mrs. S. Whitemore, wife of Mr. Whitemore, of Apsley, in Shropshire. She was about thirty years before considered to be the handsomest person in England. Littleton recalled seeing her at Bridgnorth races in 1810, when she was sitting on the box of a barouche by the side of her husband. So striking was her beauty that she was much more the object of attraction than the proceedings on the course.

At the time of Queen Victoria's Coronation, Littleton called on the Duchess of Sutherland, who had just returned from Paris, bringing the latest fashions with her. "I am always charmed in her society, to say nothing of her beauty," said Littleton, and he proceeded to describe in detail her ladyship's velvet bodice, her petticoat, and her gown of white muslin.

"The prettiest girl I ever beheld. Tall, rather large, a gracious head and face, a perfectly faultless figure," was his estimate of a Miss Stuart, whom he saw in Scotland in 1839.

Among a party at Teddesley on one occasion was "very pretty Mrs. Locke, 22. She is very fair and delicate, wonderfully pretty and of exquisite figure, very lively and amiable." She was a cousin of Lady Wrottesley.

"She stains her lashes and the rims of her eyelids," is his comment about a Miss Knightley, and the arched eyebrows of another lady so impressed him that he drew a sketch of them in his Journal.

When he first "saw a lady skait," it was on Teddesley Pool in December, 1846, and he does not forget to add that she had an exquisite figure.

He went to Brewood one afternoon to call on Mrs. Ledbury, in the same house in which he had admired her when he was a boy at school thirty-four years before. Another day he walked down to Penkridge to see the new postmistress, Miss Scarlett, and he was charmed to find she was dressed in silks, ringlets and curls as if she were going to the opera.

It is not surprising to find that sometimes he had to be very cautious in what he said to the ladies, and that on one occasion he found it necessary to call again on Lady Manners to caution her not to repeat something he had told her the previous day.

Beauty and large feet were among the topics of conversation with the Marquis of Anglesey thirteen years after Waterloo. "I was saying to him that I had rarely known a really fine woman with a small foot. He quoted me the Duchess of Richmond, his daughter, of whom he spoke justly in terms of great admiration, notwithstanding that she had nine children."

So says Littleton in his Journal, and it was fortunate for him that the old Field-Marshal did not follow the example of the woman on whom Littleton called one day, and who detained him while she told him the full story of the births and lives of the sixteen children she had reared out of the nineteen she had had.

On Hyacinthe's fiftieth birthday, Littleton wrote in his Journal: "Never did a man make a happier union than mine has been." Years later he records the inestimable blessing he had enjoyed of a wife whom he had never ceased to regard as an angel lent him by Heaven to be his guide. On one anniversary of "the most eventful day" he gave Hyacinthe "a new shawl and she made some coat buttons for me."

In their silver wedding year, Littleton "made Hyacinthe get out of bed in a hurry by telling her there was a mouse in bed. April Fool's Day!"

I RAISE MY HAT ON MILFORD HILLS

Milford Hills and Milford Hollow provided me with a contrast of two extremes. Standing on the summit of Spring Hill, I raised my hat to the memory of Thomas Anson, who, one hundred and fifty years before, caused the groups of Scotch firs to be planted on the tops of the Cannock Chase hills around me.

From the heights I made my way to the dips and undulations of The Hole or Hollow, where I struck my head against a low door frame of one of the rock cottages. It was my own fault that I had a dazing reminder that I was not spreading myself in a sentimental salute on Spring Hill.

As I was about to go into the larder, which is hewn out of the solid sandstone, my attention was diverted for a second or two by a beautiful silver cup which was standing on the dresser in the little kitchen. It was then my curiosity and my head had a sharp rebuff.

Having adopted the four-feet attitude of a coal-pit I went through the doorway, rubbing my head, and indulging meanwhile in a comment about the silver cup.

"It belonged to a young professor at Owen's College, Manchester, who was killed while mountaineering on Scafell. His father was

Mr. Marshall, one of the founders of the Midland Institute, Birmingham, and he gave it to me as a souvenir when I left his service after being head cook there for a number of years." Such was the information given me by Miss Evans, the sole occupant of "Rock Cottage," which nestles in a chiselled cavity of Sleepy Hollow, where quaint, hermit-like dwellings hide in a dozen dips and burrows, and there are gardens of the daintiest cosiness with paths as steep as the sides of houses.

Squatters made crude habitations here a hundred-and-fifty years ago, and they held fast to their encroachments in face of sundry devices—including free potations at the Barley Mow— to dislodge them. Bedrooms have been made airier, chimneys have been given less smoky routes, and tap-water has taken the place of intermittent springs. The two-ton slab of sandstone which makes a cool berth for the food is still one of the unchanged features shown to visitors who are wooed into the dell by the notice on the wicket gate: "Teas provided."

Milford has the stiffest hills of Cannock Chase, and it is there that youthful climbers can show their stamina, and wave a peak-top hail to their elders who have faded out when the valley levels begin to incline. Walking first on one hill and then another I stood beneath the gaunt arms of the sentinel trees and looked out to where Sherbrooke and

Shugborough spread their groves. Along the white curving ribbon of road that touched the bounds of valley and park J. B. Priestley passed when, as he tells in his "English Journey," "between Stafford and Rugeley I came to heath country, glowing with autumn, that was as pleasing as you could wish."

Banks of white and pale grey clouds were racing overhead, and there was a refreshing tang in the frisky wind. The Wrekin, as usual, was insisting on coming into contact with the Cannock Chase scene, and nearer at hand Stafford Castle let it be known that it, too, could not be shut out of the picture. Out Tixall way there were woodland sweeps secreting haunts of ancient peace.

Stafford town itself displayed an assortment of chimney stacks, which could be taken as read, and so could the multitude of red-roofed houses in which so many ardent lovers of Cannock Chase live. The Bank Holiday crowd had been in possession of the hills and the common two or three days previously, and the clouds of smoke which rose from the joint sand-pit and refuse-tip by the side of the main road proved that the litter pyre was being refuelled.

Roundabouts and swing-boats were being dismantled on the green; a couple of showmen's caravans moved off in the direction of Rugeley to the accompaniment of last-minute shouts and "Don't forget" parting instructions to those left behind. Leaving the green I walked up the path that passes the Sister Dora Convalescent Home, recalling, as I looked into the garden, the pre-war years when I tramped over from Hednesford to the annual meetings of the Home, and had pleasant sun-set saunters back over the moors after tea had been enjoyed beneath the bowers of rambler roses.

I went once more by The Hollow way, skirted Oat Hill, went over Coppice Hill, and dropped down into Sherbrooke.

A JUBILEE AND A DISTANT VOICE

"Daddy" Partington had only one leg, but he possessed something the like of which no one else on Cannock Chase could boast : a blue silk waistcoat.

He went so far as to claim that there was not another waistcoat equal to it in the whole of England, for when he was a youth of fourteen at Middleton, Lancashire, he wound the silk on the bobbins, and his father wove it on a hand-loom. When John was preparing to get married at the age of twenty-four he had the piece of silk made into that waistcoat, which had been one of his best treasures ever since.

On his wedding-day he wore the blue silk waistcoat and a frock coat; he put it carefully away afterwards and during the many years he had preserved it he had worn it only on special occasions.

When one morning I saw "Daddy" in his wedding attire passing along the street where he lived, I knew he must be celebrating some anniversary. It was his golden wedding day. His breast-plate of blue and his mail of black were as spick and span as if they had just come out of a band-box. His long hair and flowing beard were white and glossy, and he had a sparkling smile.

When we drew near to each other "Daddy" raised in the air the thick, heavy stick which had been his walking friend for more than fifty years, and said : "Here I am again in the dear old waistcoat, but it's more cosy and precious to-day, for it's Jubilee Day at our house."

Then "Daddy" began to call back old times, and he did not forget to tell me that the Lancashire "dear" who had looked after him and the blue silk waistcoat for fifty years was the best wife in old England.

"What do you think I did this morning before I came down the street to celebrate the wedding?", asked "Daddy."

"Why, I went on to the backyard where the ground was a bit wet and soft and I just made as many little holes in the ground with my 'peg-leg' as years I have been married, and I said to the wife : 'There, every one of them's a year as I've had thee, and God Bless thee, everyone has 'bin a good 'un.' "

. . . . .

Walking one Sunday afternoon over the fields from Blackfords towards Cannock, I was puzzled by the murmur of a distant voice. Before I could see the tower of the old parish church the strange sounds came to me over Calvan's Hill. I reached the edge of the ridge and looked round for the owner of the voice, but the meadows were deserted, and there was no one on path or stiles.

Every few yards that I went forward along the field path the voice grew more distinct, and when, several minutes later, the church came into view I scanned the enclosure within the pinnacles to see if the man of the voice might be there. The tower, like the meadows, was void.

I leaned on a stile and listened intently. I could distinguish the word 'Sinner' carried on the breeze. 'Sinner' I heard yet again and again; all else was indistinguishable.

As I continued my way towards the town the distant voice increased its volume of strange pathos. Turning at last into the Market Place I saw a man in simple dress standing in the centre of the square. He was addressing no group of listeners, but only the passers-by, few of whom troubled to turn their heads towards the speaker or to hearken to his words.

Now and again he gave a passionate fulness to his voice as if determined that his words should pursue the heedless.

That was why I had heard the mysterious sounds three-quarters of a mile away. He it was who had been saying, "Sinner," "Sinner." He used the word several times again as I stood on the edge of the footpath by the Crown Hotel and listened to him.

He was standing with his back against a guide-post. There were four hands pointing to Stafford, Wolverhampton, Walsall and

Penkridge, and stretched forth half in anguish, half in love, were the hands of the wayside preacher.

One hand seemed to point to heaven, and one was as if pointing to hell, and still did he with loud voice appeal to the "Sinner," "Sinner," as did George Fox in the snow-covered Market Place at Lichfield more than two hundred years before.

Thus did I first meet with "Johnny-come-from-God," as he was called, who, like a prophet of old, often took his stand in the heart of the busy town and proclaimed his burning message.

HUNTSMEN AT HEDNESFORD FINGER POST

One of the favourite meeting-places for huntsmen in the early years of the nineteenth century was "Hednesford Finger-Post," and a record of the period states that at "Hedgeford Hills, the centre of the Chase," was a spot where friends converging from Stafford and Lichfield used to arrange to meet when purposing to have a day's sport on the moors. Hednesford's particular geographical handiness as a jumping-off ground for the Chase has been accentuated by the railway authorities, who have given the town the added distinction in time-tables and on station platforms of "Hednesford for Cannock Chase."

The B.B.C. has informed the world of listeners that "Hedgford" is an optional pronunciation of Hednesford, and it is quite true that some of the aged natives and a few people who are subject to odd infections indulge in the dialect version. "Hensford," however, is the pronunciation which is taught in the schools of the area.

Riding on horseback over the Chase in the middle of the eighteenth century John Wesley found only Hednesford Pool where now is the mining town. Hednesford as it is to-day did not exist when Edmund Peel, third son of the first Sir Robert Peel, built a hunting lodge on the edge of the twenty-seven acre sheet of water in 1831. There was open country all round the lodge, and scores of deer could often be seen slaking their thirst at the lake.

Edmund Peel was born in 1791, and lived at Fazeley. At the age of twenty-one he married the second daughter of John Swinfen, of Swinfen. He was for some years member of Parliament for Newcastle-under-Lyme, and was an hour or two too late to be nominated at Wolverhampton for the seat vacated in South Staffordshire by the newly-made Lord Hatherton in 1835. Peel was a turf enthusiast as well as a keen huntsman, and Hednesford attracted him because it was a great centre for training racehorses, there being one hundred and twenty racers in the care of a dozen of the leading trainers and jockeys.

Samuel Lord was Peel's chief trainer. He achieved a long list of successes for his Fazeley patron. Near to the south door of Cannock Parish Church may be seen the tombstone of "Samuel Lord, trainer, late of Hednesford, who died January 11th, 1857, aged 57 years."

Peel was one of the trustees at Lichfield races, and he had as one of his colleagues General William Dyott, of Freeford Hall. In August, 1840, when the General was in his eightieth year, he "ran over to Hednesford" to see Peel because he was not satisfied with the arrangements for the forthcoming Lichfield races. He found Peel in bed suffering from gout, a complaint which Dyott had no hesitation in declaring was entirely due to his friend's incredible indulgences in the pleasures of the table.

Dyott was much struck by Peel's "training lodge, a remarkably snug box on Cannock Chase," but was sorry to learn that within ten years of the erection of the residence and stables the owner had reduced his stud considerably. The explanation was that Peel had turned his attention to "training black-faced sheep," of which there were thousands roaming about the hills surrounding the lodge.

Peel's racehorses and sheep all disappeared after a few years, the gout playing havoc with good intentions, and the lodge remained empty for some time. The stables were eventually reopened by Eskrett, who had a successful career there as a trainer. Later, he moved to the south of England to take charge of other stables. A few years afterwards, when the lodge blossomed into the Anglesey licensed house, Eskrett returned to the town to be master of the inn and the stables.

Although the Anglesey has been an hotel for many years it is still called "The Hall" by many of the older inhabitants of the town.

On its Lawn have been staged most of the town's public ceremonies and public assemblies. It has been there that Kings of England have been proclaimed to the inhabitants, and when hospital carnival kings and queens have paraded and made merry the Lawn has been the scene of the festivities.

In the middle of the nineteenth century it was no uncommon event for ferocious fisticuff encounters to take place on the Anglesey turf in the full glare of publicity. Hefty miners, stripped to the waist and with bare-knuckle weapons, preferred to settle their grievances out of court, and the iron railings along the front of the Lawn made a capital enclosure for the bruisers.

One day there was immense excitement on the Lawn when there came to halt a procession, headed by an improvised band, and followed by Old Jacky Wright, the self-styled "King of Hednesford." He was lying in a coffin which he had made to suit his dimensions, and when the bearers had rested their burden Jacky sat up in the coffin and made a characteristically funny speech to the crowd.

A night or two later as a pit-sinker was going home by moonlight he was astonished to find Jacky sitting in an armchair in the middle of the road not far from the Lawn. He was brandishing a large butcher's knife, and threatening in the most blasphemous manner to cut and carve to pieces anyone who came near him.

The peak event came when "The King," in a burst of bravado, drove his horse and gig at full tilt out of the Anglesey yard, past the Lawn, through the gate and bang into the window of a shop on the opposite side of the street.

SPECTRAL SIR HUGH RECLAIMS HIS GOBLET SKULL

One Christmas Eve in the middle of the nineteenth century a merry party of men was gathered in the dining-room at Hatherton Hall. They had all been at Christ Church together, and Hatherton, the host, full of hospitable geniality, ordered the butler to bring in a bottle of the best port. It was known as the family port and had been in the cellar for twenty years.

Later, as the men were waiting for the ladies to return from a ball, they went into the study, which had a goodly stock of guns, hunting-whips, stags' antlers, fox brushes, pistols and other features of a sporting character.

"Ah, Hatherton!" said one of the guests, who had been looking about the room, "why, what have you here? On what rare occasions do you use this goblet of such questionable shape?" As he spoke he held up to view a human skull, which had been so fashioned and prepared as to answer the purpose of a drinking-cup.

Hatherton explained the head belonged once upon a time to Sir Hugh de Hathertone, an illustrious ancestor of his. The skeleton of Hugh was unearthed some years before on a spot where a private chapel used to stand. The head had become separated from the body, and so he had it lined with silver, and set up as a goblet.

Having spoken thus Hatherton filled the goblet with Burgundy, and as they drank from it they all vowed wine never tasted so well as it did out of that ghostly bowl.

"And now," said Hatherton, "that we've pledged ourselves, each other, and heaven knows what other person besides, let us drink the health of Sir Hugh who formerly owned this head."

"Wherever you may be, we drink to you, and would that we could have you with us to-night, even though we had to surrender you your head once more," said Hatherton, as he raised the goblet.

As the last stroke of twelve was fading away the drained goblet was placed on the table.

"Hush! What is that?" exclaimed one of the guests, and sounds of footfalls were heard advancing along the corridor.

The skull on the table began to move—it rolled slowly round, till, at last it reached the edge, and then it dropped off, and appeared to fall beneath the table.

The ghostly form swung into the room as the door opened of its own accord. The headless form approached the table; it bent down.

"Sir Hugh de Hathertone!", all the

startled men cried, but the skeleton knight gave no answer. He made a true knightly bow, closed the door, and strode down the corridor, the clank of his spectral armour gradually dying down as the figure receded.

The arrival of the ladies from the ball drew the terrified host and his guests from the library before they had opportunity to pursue their investigations into the extraordinary happenings, but they lost no time on the following morning in making a search for the skull goblet. It was nowhere to be found. No one had seen it since the incidents of the previous night.

Hatherton's study opened on to a lawn, and when the searchers stepped through an open window to hear the clear music of the Christmas bells at Cannock Church, they picked up from the crisp frosty grass what appeared to be a ball of silver.

It was found to be a thin silver plate rolled flat into the shape of a ball. It was the silver which had lined the skull goblet.

In vain was the search for the skull. The only conclusion Hatherton and his friends could make was that Sir Hugh de Hathertone, while indulging in the pardonable wish to regain property which was without question legally his own, had a knightly horror of taking what by the most rigid interpretation of the law he could not with propriety claim.

Such is the story of the Hatherton skull goblet as gleaned from an old book.

DIM CENTURIES IN COURT BANK COVERT

It was the first time I had been to Court Bank Covert at the end of May, and I did not expect to find such great wealth of bluebells as came into view as soon as I strode away from the Cannock Wood road and reached the edge of the grove. My previous visits had been made when luxuriant bracken brushed about my shoulders as I forced my way through the thicket.

My reason for going to the covert this time was to see how far the rumour was true that Nun's Well was in danger of collapse. A few bricks had been displaced, but there was little else to be anxious about. The ancient oak still made a sheltering bower for the well, and its roots had a python-like hold of the surrounding rock, boulders and ironstone slag that buttressed the dim reservoir where a small, unruffled saucer of water lay. No rain had fallen for several days, there was no dripping or gurgling of water within the well, and there was a stagnant pocket or two in the parched beds of the one-time rivulets.

A few yards away the Redmoor Brook was chattering in its deep-set, narrow trench, where it hurried towards the moat within whose bounds the King's House was formerly situated. The site of the mediæval house was well defined, the moat retaining its formal rectangular lines. There were no traces of any building, and oaks which stood in the enclosure had prospered through several centuries.

Many of the oaks in the covert provided ample material for a study of singular attitudes and quaint contortions. A veteran with a girth of twenty-two feet had a kingly pose, and just where it could be expected to have come to an end of its productiveness it thrust out for another eighty feet overhead a new extension. A massive neighbour rose twenty-five feet from the ground before it ventured on secondary branches, each thick enough to be a tree in itself.

A rival of the eighty-feet stalwart had lumps of smelting-furnace slag wedged into the folds of its stanchions and its hide was like that of a rhinoceros. There was little doubt the tree was standing where it had already its foundations when the iron-makers of Elizabethan days worked there. One extraordinary oak was hollow for the first twenty feet of its rise, and it seemed incredible that from such a timber tripod wreckage there could emerge another forty feet of growth whose topmost arms gave greetings to the centuries. There was a derelict oak, long since past its green leaves stage, which had a fascinating poise which reminded me of Eros perched in Piccadilly.

A keen competitor for oddity of gesture was an oak which, from a bare base of thirty feet in height, suddenly pierced the canopy of the covert with a sixty feet long lance at an angle of forty to fifty degrees.

The Pillar Box Oak had a shapely and beautifully bevelled letter-hole, placed about three feet from the ground, and it was so realistic in appearance that I posted a single-sheet missive as an accompaniment to the fancy. The Elephant Oak was black, dwarfed and bulky with its hundreds of seasons. It had long ceased to respond to the call of springtime, and the birds ignored it as if it were haunted. One gaunt branch was pushed upward, just as Jumbo would strive with his trunk to reach something overhead.

From the oaks I passed to the neolithic flint chipping floor or factory, the only one of its kind in Staffordshire. Instead of thinking in mediæval hundreds I began to ponder upon prehistoric thousands of years. In this Redmoor, or Radmore, Valley the Iron and Stone Ages had left their evidences, yet not one man in a thousand on Cannock Chase had searched here to try and read the story of the past. Within a few yards of Henry the Second's moat flints have been found in an area of nearly two acres of clearing within the bounds of the covert.

Formerly the land was wild and un-cultivated, and it is since it has been tilled that the site of the chipping floor has been revealed. About twenty-five years ago Mr. G. M. Cockin, of Rugeley, and Mr. T. C. Cantrill, who were Fellows of the Geological Society, found about six hundred specimens of flints, including nearly thirty implements, on this land, and the conclusion at which they arrived was that Court Bank Covert was a spot where flint-making was carried on.

A large quantity of scrap flint has been left scattered over the place by the prehistoric craftsmen, who, apparently, had obtained their material from outside sources, there being no trace of flint in the local strata.

In 1862 a flint "strike-a-light" was found on Castle Ring, which is only a short distance from Court Bank Covert, and in 1908 there was picked up on the Ring a flint scraper of unusual formation which was considered to be coeval with the prehistoric occupation of the Ring earthworks. These facts open out a vast field of surmise with regard to the British-Roman encampment having a far greater antiquity than has usually been accepted. Flints are still found on the site of the chipping floor, the best time for investigation being when ploughing has been recently done, and there is a possibility of new specimens coming to the top.

Zeal for flint specimens did not overcome my consideration for the crops of Robert Bailey, whose cottage stands on the fringe of the covert, and when I went inside the homestead in which a dozen generations of Baileys have been born, I was given several good flints which had been found on the chipping floor.

Tales of grandfathers and great-grandfathers flowed into our conversation, and times were called to mind when the cottage of the Baileys was innocently forfeited to the Old Marquis by the payment to him, first of a penny, and then later of a shilling a year as a courtesy acknowledgment of right of possession. This developed into a rental payment, and years had to go by before the Baileys felt safe and sound under their family roof-tree.

Thus in the course of a May afternoon did I touch the centuries at many points, and look through differing windows with memorable vistas.

AMONG THE BOOKS AT WYRLEY GROVE

Within a few minutes walk of the stretch of Watling Street which lies between the Fleur-de-Lys and the Old Turf inns, there stands Little Wyrley Hall, or Wyrley Grove, personal choice governing the use of the alternative names.

Pit-shafts and spoil banks are features of the landscape not many yards away, but the mansion is screened by an imposing bower of trees, and the ancient and the modern seem centuries apart.

Little Wyrley manor was part of the Bishop's holding in the time of William the Conqueror. It passed through various hands in later centuries, including John Leveson, of Wolverhampton, and Roger Fowke, of Brewood. From the last-named the lands went by marriage to the Hussey family, who also became owners of a third part of the manor of Norton Canes and a considerable area of freeholds in the parish of Pelsall.

Wyrley Grove had its beginnings in the seventeenth century. It was soundly built of fine English brick and sturdy oak. Alterations were made, particularly on the south side, in the days of William and Mary, and at one period there was a large conservatory which had close contact with the library. This unattractive glass structure has been demolished in recent years, and the external appearance of the house has been much improved thereby.

At heart, the mansion retains a great part of the original building, and accretions which were introduced according to the whims of the successive owners of the house have been subjected to careful and effective restoration by Mr. and Mrs. Frank Wallace, who are honourably maintaining the Hussey and Macpherson traditions.

The coat of arms of the Fowkes is prominently displayed on the external walls of the hall, and windows tell the heraldic story in bold colours and historical detail. At the threshold there are to be seen immense beams hewn from Cannock Chase monarchs of centuries ago. An oak staircase of exceptional beauty of design dominates the entrance hall, and nearby is a magnificent chest, carved and figured within and without. Chairs of the Cromwellian period, pewter dishes of super-circumference, handsome doors with heavy metal-work clamping them, and a sociable fire set beneath a roomy arch all help in the pleasing association with the historical past.

On a previous occasion when I visited Wyrley Grove a cupola rose above the roof, and its weathercock spindle operated a golden arrow which revolved on a dial fixed in the entrance hall of the mansion. By this means it was possible to learn of the direction of the wind without the trouble of going out of doors. This curious device had been partly dismantled, but the dial and arrow were left as reminders of the heirloom equipment.

As I went through the hall I remembered how Miss Mary Bagot called there in November, 1827, and was impressed by it because "such abodes are really becoming invaluable from their rarity, and as specimens of the olden time." The feeling she had in company with the Husseys there was that she was being wafted back for a hundred years. The heir apparent, a rough enormous boy of six, was the first who had been born in the direct line for a hundred years, and was prized and spoiled accordingly. On that particular day he had a cold, and Miss Bagot saw the mother doctor the precious lad with syrup of violets.

Phineas Hussey, of Wyrley Grove, died six years after the date of Miss Bagot's visit to the hall, and he is commemorated by a tablet in Norton Canes Church. He was a keen collector of prints, and he placed his treasures in rooms known as the Florence and Orleans Gallery at Wyrley Grove. A century after his death two engravings from the collection were sent to Sotheby's salerooms, London, and they sold for £545. Of that total, £400 was obtained for a first impression of Rembrandt's "Dr. Faustus," its special value being in the fact that it was without an additional shading which was imposed on later specimens.

As soon as I entered the library I fancied I saw Phineas Hussey there among his favourite books. On the shelves were hundreds of volumes which were first published between two and three centuries before, and it was possible by a glance at the ranks of time-worn tomes to form an idea of the reading tastes of the gentleman of leisure who formerly ruled there.

The Annual Register was there in successive rows over a period of at least two hundred years. The first volume of the Gentleman's Magazine, dated 1791, and a numerous progeny for over a century rested in seclusion on a top row. I reached down one of the numbers for 1799 and read once more the Latin inscription which a correspondent stated he had seen on the gateway of Wyrley Grove, the date 1687 also being given. "As to you, so to others" is a free translation of the inscription, which is still to be seen.

A Universal History of the world, many works of famous historians in numerous small volumes, a large number of books of travel in all parts of the world, a profusion of old plays, including Hollinshed's "Chronicles," rubbed shoulders with a quantity of bygone theology. Early English novels kept company with an assortment of poetry of the eighteenth and early nineteenth centuries.

As was to be expected, Plott's "History of Staffordshire," dated 1686, was there, and there were evidences that "P. Hussey, Esq.," whose name was written on the inside cover, had read the book often. There were striking illustrations in a 1728 edition of Bunyan's "Pilgrim's Progress," and 1765 was the date of a twelve volume edition of Dean Swift's works.

A first edition of Sterne's Letters in three volumes, dedicated to David Garrick, was dated 1775. A 1793 edition of Boswell's "Johnson" had been well-worn; Johnson's Dictionary in a miniature 1779 edition had novel features, and four volumes of the Letters of Lady Mary Montague had a special interest by reason of her local associations.

JOSEPH THE DREAMER

It was on the occasion of the Old Hednesford pit disaster when I first met Joseph the Dreamer, that being several days before the Christmas of 1911. Five men lost their lives by a fire which broke out at the bottom of the shaft, and there were many deeds of daring done before the flames could be overcome. For bleak raw days and bitterly cold nights watchers waited patiently at the pithead to learn the latest news of happenings belowground. From time to time I eased the dragging hours by sheltering in the little office where Joseph was keeping a glowing fire at full radiance, and maintaining a continuous supply of cups and jugs of hot tea and coffee for the toilers and the watchers. He also supervised the sandwich department.

His usual duties concerned ledgers and daybooks, sales registers and weigh tickets, during specified hours of the day, but ordinary restrictions of peaceful routine were set aside in such a time of tragedy and need, and Joseph of the soft voice, the gentle bearing, the imperturbable outlook worked without stint and with words of good cheer from nightfall to break of day and far beyond.

When at the end of a long vigil there was brought up the shaft a hastily prepared simple coffin on which some comrade of the mine had chalked in bold capitals, "TOM STOKES, HERO," Joseph shepherded us all into a group, and, with closed eyes, and raised hands he said, "Thank God for a brave man; God bless the widow," and back he went to the making of tea and coffee.

In later days, I called many times to see Joseph in his office. It was at no garnished and meticulously tidy desk that Joseph did his figuring. His books rested on cushions of papers which some day he hoped to be able to put into their proper files and pigeon-holes. There were hundreds of them covering the desk from end to end, but he seemed able to find any one of them at the right moment. The littered desk provided him with evidence to demonstrate his powers of location.

One afternoon when I leaned on the desk to make a few notes of things he was telling me about Cannock Chase incidents of the past, Joseph looked over his spectacles with instant concern and demurely said: "Now mind you don't upset any of my papers, else I shan't know where to find them."

Joseph was proud of his memory, and to prove that his claim was not unfounded he opened a drawer and produced from it a cutting of several years before of a "Pitman" account of some Cannock Chase feature. At his request I held the half-column article while he repeated word for word without a single mistake the particular historical facts which had caught his fancy and clung to his memory.

When Joseph talked of Cannock Chase it was chiefly of churches and schools, vicars and curates, wardens and schoolmasters whom he had known over a period of fifty to sixty years. Most of his best thoughts were of the Church on the Hill at Hednesford. There was not the least doubt that when he was turning over the pages of his colliery books and totalling the columns of figures there were mingled with his thoughts anxious considerations for St. Peter's. The only money of which he had real concern was that needed for the church of his fathers. When the offertories fell below the desired average he would recite the numbers of the coins of the various denominations, and place a profound emphasis on those of the lowest value.

Joseph had been deputy organist for many years, he had served as a warden, and he knew every nook and cranny of the building better than vicar or verger. So concerned was he about the safety of the church that every night for many years he toiled up the steep hill from his home at Littleworth to make sure that all was well, and to lock the doors when darkness fell.

Passing by the church one winter night when the snow was on the ground, I heard the sound of closing doors. I looked towards the south door and there was Joseph happy in the thought that once more he had put mother church safely to rest for the night.

I waited a moment or two for Joseph, and when he came through the gates he still had his key in his hand. It was much smaller than I had expected it to be, and it was polished like silver. He had worn it bright by always carrying it in his pocket.

Joseph the Dreamer, surnamed Lindop, is commemorated in the church by an oak screen placed there by his two brothers—one a Bachelor of Arts and the other a Bachelor of Music.

JAMES TURNER TELLS OF THE "POCKET VENUS"

When James Turner celebrated his Diamond Wedding at the age of eighty-three, I went to see him in his cottage on Hednesford Hills, within a few yards of the spot where cock-fighters and prize-fighters used to crowd into a sheltered hollow.

James was born three years before the "Waterloo" Marquis of Anglesey died, and as his father—John Turner—was in the service of the Marquis, the lad spent his early days in surroundings where tales of the great soldier were many and oft-told.

It was in a thatched cottage at the foot of Castle Ring, and near to the Rawnpike Oak, that James Turner "came to town," as he called it.

"Seventy-five years ago I used to get into the big hollow in the middle of the oak, and now it is gone, but last Sunday I spent a few minutes sitting on the blackened stump," said he. In the same cottage where he was born Methodist preachers used to hold forth while the dumplings were wobbling and sputtering in a pot on the cottage fire. When the thatched cottage became dilapidated the Turners went to live for a time in the Grand Lodge at Beaudesert Hall.

At the age of eight James often carried

ammunition for the guns as the Pagets and their almost continuous houseful of guests went shooting in New Hayes Wood and Court Bank Covert, to mention only two of the best-known spots where bags could be easily filled. He had seen them shoot seven hundred rabbits, two hundred pheasants, and quite a number of snipe and woodcock in one visit to the New Hayes. The Pagets were dead shots, and they never seemed to get tired of their sport.

There were numerous great sporting days when noted people would be there, and the beaters and the lads who did all sorts of odd jobs were never short of money in their pockets. Lord Henry Paget, the second Marquis, was very generous, and Henry William, Lord Uxbridge, Lord Alexander, Lord Alfred Henry, Lord Charles Edward, and Lady Florence Paget were all good friends to the people on the estate. It was always a treat to go out with them into the coverts.

Lady Florence was the only daughter of the second Marquis, and because of her remarkable beauty was known as the "Pocket Venus." She eloped with Lord Hastings on the eve of what was to have been the wedding of the season, Henry Chaplin, the bride-groom-to-be, being jilted at the last moment. She put on her wedding dress and showed it to her father, and then threw it aside, and went off with Hastings.

Lady Florence had a favourite swan at Beaudesert, and one day the bird caused a sensation when it dared to attack a famous lady of title who was a guest at the mansion.

"There is a story which is worth telling about Lord Berkeley Paget," said the old man. "When the Marchioness was expecting the arrival of her baby, there were very anxious times and it was feared that either mother or baby would have to be sacrificed. The doctors pooh-poohed the idea that a village midwife's offer should be entertained, but the situation became so grave that Mrs. Stringer was fetched to the hall. She made the condition that no one should be in the room with her, and Lord Berkeley Paget arrived safe and sound. From that day Mrs. Stringer was a worshipped woman in the neighbourhood, and she always seemed to go about with a feather in her cap.

"And it was Mrs. Stringer who brought me into the world as well," added James.

Beef, soup, mutton and dripping: what fine servings were dispensed on stated days at the hall to the Paget's workpeople. Turner, senr., had duties to discharge in the kitchen at the hall, and many times he skipped home with a tablecloth-full of odds and ends sent back from the dining table in the Great Hall. Having deposited his load on the cottage table he would hurry back to the kitchen.

"Have a horn of beer, my lad?" was the question put to James when he went to the hall to see his father, and not once did he say "no." The children of the workmen were allowed to gather the "leesings" of the corn left on the fields, and it was no uncommon thing for enough to be gathered to make three stone of flour, this being sufficient for a good baking of bread.

James took the corn to Gentleshaw Windmill, and he spent hours waiting for his mother's corn to be ground, the flour at length being placed in one bag and the bran for the fowl in another.

The old man could not resist telling a little yarn about the Windmill. It concerned the Old Marquis and the keeper of the Windmill Tavern, who had made up his mind that he would not become subject to the Paget overlordship. He outwitted all efforts made to win him over.

One day the Marquis went to the Windmill Tavern, and said, "They tell me, Bonehill, that you won't pay any acknowledgment to me."

"Well, they speak the truth, my lord; good morning," was Bonehill's rejoinder, and the interview came to a speedy end.

Joseph Richards was the keeper at the Park Gate Inn, and he too was noted for his plain speech and his talent for maintaining the balance between the Bible and beer. When a preacher went on too long with his discourse Joseph had no hesitation in walking out of the chapel and going off to open the Park Gate for customers.

"HALF-SOAKED DICK" ON MUSIC

"I don't know how old that piece is and who made it, but it's a top-hole bit of music," was the comment of "Half-Soaked Dick" at the conclusion of a choir's singing of "A Grecian Landscape" at Cannock Flower Show.

Sitting at his table a few feet away the adjudicator heard Dick's words, and quoted them in his review and announcement of awards. Dick felt as pleased as if he had won a prize, and was prouder still when he found that folk were saying to each other: "Did you hear the judge mention what 'Half-Soaked Dick' said?"

Dick was always in the front rows of the singing tents or band enclosure at the flower show. He was a broad-built, comfortable looking fellow, whose unpretentiousness of attire included a red-and-black scarf which established his identity many yards away. He would listen through a long programme without budging an inch, and although he seldom troubled to consult a list of items he always seemed to be making mental comparisons and coming to conclusions as he went along.

"Let's get as near as we can to 'Half-Soaked Dick' so that we can hear what he has to say about the music and the singers'' was often what people said when going to the choral concerts. They knew that most likely he would indulge in a critical commentary in quaint and entertaining phrases, punctuated with original similes.

"That's almost as good as our best lark on Hedgeford Hills,'' remarked Dick when a smiling soprano ended her trilling, "but the lady nearly lost her footing near the top of the ladder.''

"My eye, I thought the double bass and the drum had tumbled down the Valley pit,'' he exclaimed, when a band snapped off its noisy climax.

"I could run upstairs to bed easier than he climbed up there, and it's time he spent a bob on a new pair of bellows,'' was Dick's criticism of a piping tenor who told him all about "Nirvana.''

"I like to hear the first violin going gently off to sleep when he almost puts his nose on the strings, and it works me up fine to see the cornet player blowing and working his fingers until his eyes almost tumble out of his face,'' said Dick on one occasion.

The hectic competition of a tenor and a baritone to put the finishing touches to "Excelsior'' drew from Dick the remark that it reminded him of the miners' agent and the boss of the local coal-owners arguing about an increase in the men's wages.

"Let's hear that 'Sallydamer' or whatever it is, again," was the request he made to the conductor of the military band at Cannock Show. He said he liked it a deuce of a lot, and Elgar's "Salut d'amour" was given as an encore to please Dick.

One day I met Dick in the company of several of his pals, and he felt he must be content with the dialect of the pit, and not bother about trying to talk specially "polite."

"I seed yer at the 'Messiah' last night," he said. "it was a darned good show. Chorus wornt too jealous o' one another. They wornt too heavy on, either front or back, although the basses gid it beans in some places. Wunce the kire got a bit messed up when they were going over the top of one of the choruses, but that theere Darby pulled 'em together fine, he did, and he deserved a military medal for it."

Dick was delighted with the bass soloist— W. Hone Boot, a local product— and said, "it was well wuth the money to hear 'Our Own Boot.' He know'd how to mek the nations rage, that he did."

Dick walked hundreds of miles to hear bands and to be present at first-class concerts. He told me how he walked to "Hampton" to hear Madame Patti sing in the Exchange Hall. He was so determined to hear her that he pawned his best waistcoat for half-a-crown and set out to walk the eleven miles feeling sure he would meet with someone who would give him a lift on the road.

It was not his lucky day, however, and when he arrived at the end of his journey he was tired out. He paid two shillings for a seat, and had a couple of drinks with the remaining sixpence. After the concert, which he declared to be a "real topper," he tramped back to Hednesford feeling very empty inside one part of him, but very pleased in another part. For days he was telling everybody in the pit and in his favourite working men's club about "Angelina Patti."

It was said of Dick that he knew where to place all the local colliery buzzers and locomotive whistles in their correct position on the musical scale. Returning from the pit he called at the Colliers Arms, and he was challenged to tap all the glasses on the counter with a small stick and then see if they made a tune of any sort.

His answer was: "I tell thee what I'll do, I'll hit thee on the yed with this hedge-stick of mine, and I can tell thee the sound will prove it's as empty of brains as my pit bottle is of beer."

George Barker, alto, Bill Hill, tenor. Hawkins, the schoolmaster bass, "Peg-leg" Froysell, who also was a champion on the deep notes, were names that set Dick rattling away in reminiscences, and Sam Cotterell, Tom Ward, Stephen Gilbert, George Rowe and "Siah" Whitehouse—the last-named earning the prefix because of his great fondness for the "Messiah"— were instrumentalists of whom he never tired of talking.

Dick lived long enough to be able to sit by his own fireside and listen to great music "on the wireless." He was too old to walk to "Hampton" and back, and was glad there was no need for him to pawn his waistcoat in order to hear a good singer.

"I wish I could see their faces when they are singing and playing like I used to; and there is no chance of asking them to give one of my old favourites as an encore. I tell you what I will do; I'll get my grand-daughter to send a letter to Brummagem and ask them to do me just a little favour by playing one of these nights several of my old favourites. But, you know, I can't tell them all their proper names, and I don't know how to spell them, but then we shall manage between me and the grand-daughter."

I gathered that "William Tell," "Poet and Peasant," "Barber of Seville," "Cavalleria Rusticana," "Il Bacio," "Ballet Egyptian," "Merrie England" and Rachmaninoff's "Prelude," were some of the pieces Dick wished to have on his own request night.

A JUNE DAY ON ESSEX BRIDGE

No day can be too dull nor too uninviting for a walk over Cannock Chase by way of Penkridge Bank, South Street, and across the Trent at Weetman's Bridge.

I found the journey a great delight, and when I turned on to the towing path of the Trent and Mersey canal at Colwich I stood for a few moments and surveyed the highland tracks by which I had come. The river and the canal run parallel towards Great Haywood, and there were many countryside beauties in that mile and a half of easy walking. Through the open door of a cottage I heard the clear sounds of a hymn being sung at a morning broadcast service.

I came to a stone which told me that if I continued 55 miles further I should come to Preston Brook, Warrington, and that thirty-seven miles at my back was the Derbyshire town of Shardlow.

Soon Shugborough Hall, gleaming white like a monster wedding-cake, came into view through a long avenue of trees, and a few minutes later I was on Essex Bridge, which, according to traditional records, was built by one of the Earls of Essex that he might the more readily get his horses and hounds on to Cannock Chase from Chartley Castle.

THE ESSEX BRIDGE.

Over the narrow pack-horse width of the bridge there came from the village an old man, holding the hand of a two-year-old child. He was blind, and was taking his grand-daughter over the river "to see the lovely fields." They went like that every day, and, said he, "The little darling, whose face I have never seen, is a rare walker."

I sat on one of the twenty-six surviving chevron-shaped alcoves of the bridge, and soon I had other company. A thrush with flush-fresh feathers settled within a yard of me, gave twitchy salutations with his head, and challenged me to look into his fiery eyes. Within the next few minutes two bullfinches made their obeisances, and welcomed me with "sweety" vocabulary to their observation promenade, Judged by the patriotic character of their plumage colours the birds might have just come from a jubilee celebration.

When a robin slipped down from a neighbouring tree and stood between a couple of blackbirds and began to help to clear the narrow stone table of any crumb or worm tit-bit that might be in the crevices, I felt I was indeed consorting to an uncommon extent with the birds of the Chase. I sprang to my feet and gazed rapturously when a kingfisher came joyously down the river, shot through an arch of the bridge and skirted the Haywood Bank like an iridescent arrow. Not until the park ornamental bridge had been passed did the kingfisher wheel round into a return drive. Once more the flash of brilliance went between the face of the waters and an archway of grey stone, and the magnificent visitant hastened up the track of the river, where rhododendrons made gorgeous the curving banks.

Birds were constantly wheeling over the bridge, and crossing from bank to bank of the river, where huge masses of trees raised their heads a hundred feet in the air.

"You see me get up there," one perky bullfinch seemed to say to me, and I watched and admired his method of mounting from bridge edge to highest branch in four distinct upward surges. A solitary swallow indulged in ten minutes of figure flying and then went elsewhere for a diversion. He came back later and staged an additional display of weaving wispy lines in space. Two swans and five cygnets kept the strict rule of the road as they sailed close to the bank of the big island which divides the Sow and the Trent, and they chose to land rather than provide a choice photographic theme in an arch frame of the bridge.

Although there had been much rain for several days, the waters were low and were making no special hurry. Their only sound was that of the lone drone of the weir, which came from silvery cascades of mere inches. In the middle of the river were small hillocks crowded with docks of super-luxuriance, their red central spikes being a yard long. Rushes resting on the surface of the river had the appearance of great recumbent plumes: some were sage green, and the pale green of others passed into copper tones. There were emerald green rushes which were so fragile that their fine points wheeled over and over, an inch below the water surface, like maddened snakes.

In mid-river a bastion weed island held a key position, and placed two of the bridge's arches almost out of commission as far as the flow of water was concerned. Fifty yards higher up the river heavy drift-wood chunks lay derelict in a plain of mud. A scuffle of wind gave the placid water in the shelter of two bridge pillars a crackle-glass appearance, and when the sun eluded the sombre clouds its rays covered the sleepy countenance of a rush-moated pool with the sheen of hammered pewter.

There was a spot where a tiny mud obstruction made the water diverge into a ladder of Norman dog-tooth ornamentations. A slender rush which measured its length on the surface aided miraculously in the creation of a striking effect, for it became like an ostrich feather with fifty fronds on each side as it met the impact of the water.

Many of the trees on the banks of the river had premature autumn tints hastened by nipping frosts. Horse chestnuts had lost most of their candles, oaks had a surprising proportion of withered leaves, and the red and white hawthorn had become shabbily faded. Scores of small leaves were constantly being released like parachutes from wide-spreading branches, and in every one of the recesses on the down-stream side of the bridge-top there was a pile of wizened wind-fall leaves.

Overhead, skies were whimsically moody: first a floating dark blanket intercepted the sunshine, and then two or three minutes later, from out of an assemblage of blue and white folds, the sun would make its reappearance. There had been rain in the early morning, and the bridge path still had little puddle reminders, but the clouds were riding high and bent on greater business than disturbing me from my Essex Bridge coign of vantage.

Up country, in the direction of Ingestre and Tixall, the view was an agricultural one; broad green meadows, cattle at pasture, white-washed farmhouses. Southward, the enchanting Chase held sway.

From this same bridge, in the summer of 1780, Thomas Pennant, making the journey from Chester to London, looked up where "the lofty front of Cannock Wood, clothed with heath, or shaded with old oaks, and scattered with glowing bloom by the free hand of nature."

Pennant found the forty-two arches of the bridge, as he had known them years before, had been much reduced in number. To-day, there are only a third of the arches left. Pennant doubted the accuracy of the local story that the bridge was originally built as a compliment to the last Devereux, Earl of Essex, of Chartley Castle, and he was convinced there must have been a bridge there long before the days of Queen Elizabeth's favourite.

Queen Bess herself passed over Essex Bridge in the summer of 1575, when she visited that favourite courtier of hers, "the good Earl Essex," honoured by Edmund Spencer in his dedication to "The Faerie Queen."

Robert Devereux, the eight-years-old son of the Earl, was artlessly unaware of the chain of events which, following his youthful enjoyments of hunting on Cannock Chase, would make him at nineteen the adored of the virgin Sovereign of fifty-three, and land him on the scaffold at the age of thirty-four.

Essex Bridge was rebuilt in 1729, several restorations have been made since that date, and in 1935 it was scheduled in the Staffordshire list of the protected ancient monuments of the county.

ALL ARTS COMBINE IN TRENT-SIDE ARCADY

The glow of rhododendrons, the song of the birds, and the peerings and scamperings of rabbits gave a touch of life to the dozy Octagon Tower, whose fluted pillars and lofty doors were contending with Father Time and enveloping ivy.

I had gone to make acquaintance with the many imitations of classic architecture which Thomas Anson had reared about Shugborough Park in commemoration of the naval exploits of his famous brother, Admiral Lord George Anson. Here was the tower of Andronicus Cyrrhestes, standing close to the drive which connects Shugborough Hall with the main road from Rugeley to Stafford. It was a striking relic of a bygone day, an evidence of pride and commemorative affection, but scarred by dilapidations, almost hidden to the passer-by behind an accumulation of shrubs, and serving no useful purpose.

Thomas Anson must have been a munificent brother, and he incurred enormous expense by engaging James Stuart — "Athenian Stuart," as he was called—to adorn the park at Shugborough with ornamental buildings, remarkable statues, and the Arch of Adrian, or "The Triumphal Arch," which for the last one-hundred and sixty-five years has held its

THE ANSON MONUMENT, SHUGBOROUGH PARK.

commanding and magnificent position on the hill beneath which railway engineers pierced their tunnel to take the Trent Valley line from London to the North.

Josiah Wedgwood tells us of serious discussions which took place at Shugborough when James Boulton's workmen found they were faced with almost insuperable tasks in trying to carry out Stuart's designs. It was a problem of the right understanding of the limitations of metals, and the extent to which Wedgwood's black basalt should be introduced.

Wedgwood was asked what he could do to ease the situation. Said Stuart: "Mr. Anson will glory in having the arts of Soho and Etruria united."

On Christmas Eve, 1770, Stuart, Boulton, Wedgwood and his partner, Thomas Bentley, and Dr. Erasmus Darwin, of Lichfield, met at Shugborough. First of all they put themselves in a good humour by indulging in "very good Christmas cheer," and then they set to work to solve their problems. They decided what Soho should do, the extent of the contributions of Etruria, and the nature of the various embellishments to be placed on the Arch of Adrian to commemorate the naval victories of the great Admiral.

Thomas Pennant made Shugborough the centre for most of his Staffordshire excursions, and he was there a few hours before

Thomas Anson's death. Pennant thought the place to be a paradise indeed, and to be so benignant in the influence it exercised that every neighbour of Thomas Anson became a partaker of the elegancies he had provided.

Adonis and Thalia were there; Trajan was in an attitude of haranguing his army. A marble group by Scheermakers represented two lovers listening to a shepherd who was reading to them an inscription on a tomb: "Et in Arcadia ego." In this way did Thomas Anson wish to remind himself that amid his delights death might snatch him away, and he is said to have devoted particular care to the preservation of this statuary.

Pennant says he found Anson "listening calmly to the melody of the harp, preparing for the momentary transit from an earthly concert to a unison with the angelic harmonies."

On a lovely summer afternoon, in the midst of such loveliness as Shugborough Park affords, I confess I did not linger long on the lugubriousness of Anson and Pennant.

As a contrast, another monument commemorates a favourite cat which was with Admiral Anson when he made his voyage round the world. The monument is surmounted by a large urn in which the ashes of the animal were placed, and a figure of the pet reclines there also.

Imposing without and noble within, Shugborough Hall is one of Staffordshire's great treasure houses. Many of the family heirlooms are associated with the great Admiral, whom Reynolds and Roubilliac have portrayed in painting and marble. The faces of many Ansons, by master painters, look down from the walls. Spacious rooms with ornate and costly furniture, ceilings whose decorative work was carried out by Italian craftsmen in the middle of the eighteenth century, and a marvellous profusion of *objets d'art* at almost every turn help to make Shugborough Hall a wonderland of beauty. The large number of Chinese works of art seen in the hall were brought from the East by Admiral Anson, and were originally accommodated in the Chinese House specially built for them in the park.

Near the hall are ivy-covered ruins which mark the site of the ancient palace of the Bishops of Lichfield, and a thousand-year old yew on the banks of the Sow is claimed to be unique in the whole of England.

It was a later Thomas Anson, who, at the beginning of the nineteenth century, in his desire to give his mansion a finer setting, removed the old village of Shugborough and re-housed the inhabitants in Great Haywood.

The present Earl and Countess of Lichfield take a great pride in Shugborough Hall and park, and in their loyalty to all good Staffordshire causes maintain worthily the high traditions of a distinguished line of Ansons.

THE CEREMONY OF THE KEY

"Does Old Harry Booker live anywhere about here?" was the question I put to a grey-haired, stooping old man who was sitting on the stone coping of the front palisading of a cottage at Littleworth.

"Quite right, this is the place; you will find my dad in the house. He's just come back from a climb up the hills," was the answer I received.

Harry Booker was 92, and was able to produce indisputable evidence of his having spent seventy years at the pits, for fifty years of which he had been concerned in the official affairs of the Miners' Union.

Harry had in his kitchen an old four-legged desk which was a survival of the days when he was not only secretary of a friendly society lodge, but also responsible as a checkweighman at the Old Hednesford pit for the records of that particular lodge of the Miners' Association. There was a mass of small memorandum books and string-tied bundles of papers in the desk, whose contents he safeguarded with extreme care. Harry had been trained in a policy of strict privacy of trade union matters, and nobody could ever accuse him of giving away secrets or letting anyone have an inkling of how negotiations were shaping behind the scenes.

He was never known to open his mouth and put his foot in it by saying a word to any "gaffers' man."

I soon learned that the four-legged desk was by no means a family affair. When Harry let his memory have free play for a while he spoke of events of the Chase seventy years previously, and said he would prove what he had told me by one of the little books in the desk.

"Don't trouble to get up from your arm-chair, your daughter can find it for you," said I, innocently, but I spoke without knowledge of the domestic regime.

"No fear of that coming off," chimed in the daughter, and her husband joined her in a ripple of laughter.

"Only me opens that desk; I take care of that, and here's the key which nobody touches but me," said Harry, as he stepped towards the desk and produced the key from his trouser pocket.

Perhaps there were many secret odds and ends concerning masters and men stored there, or there might be details about arrears of weekly pay due from hard-up miners of years before which it was not nice for other eyes to see than those of the authorised secretary of the lodge. If the desk held such documents I did not see them, but Harry brought out for inspection a list of names of men who joined the Union sixty-five years ago. Many of them came from Shropshire and the Black Country, and I recognised names of men whose families have since become well-established in the district.

There were many old minute-books with curious entries, scraps of paper on which Harry had made notes of things he wished to say at meetings of the Miners' Council, and when he turned over a page at the head of which was written the word "Blacklegs," he closed the book with a word of apology, saying "Nobody's ever seen those names but me, and it will do nobody else any good to read them."

Having spent a few minutes with the books and papers, Harry put them back again into the desk, fastened the lock, and back into his pocket the key went again.

I had seen the ceremony of the key at Harry Booker's.

"When I have charge of the affairs of any men I don't believe in letting other people know about them. My family know all about that being my principle, and I have a notion they think all the more of me for it. I know they have a laugh sometimes about the key never getting into their fingers, but there's one thing they know as well, and that is that there's no skeleton in that desk, and there are no bills that have been waiting for years to be paid."

Thus spoke Harry after he had returned to his armchair. He then asked me if I knew anything about Bloxwich. Did I know the Green, the Bull's Head, the Swan Inn? He was born at Bloxwich, where his father was a bit-maker. Harry preferred the pits to the bits, and worked down below for Rowley and Reeves near the Swan.

Later, he migrated to Hednesford, where he soon found that his job was hard, with little spare time to stretch his back and his legs. If he had a holiday he got it by stopping away from the pit for a day, for there was no other way.

"It was all right to feed yourself and the family on dripping and bilberries picked off the common during the sixteen weeks' strike," was one of Harry's asides.

There was a rare set-to in the King's Arms one night when an argument bubbled up about who was the most likely parson to be made the first vicar of Hednesford. A good, free fight soon made the place all of a 'fizz', and the landlord was almost crazy in his struggle to keep his crockery from all being smashed to bits.

Even on the day when the church was opened there was a fight between two men on the edge of the crowd outside the gates because they could not agree about the correct way to spell the name of the new vicar—Mr. Pauli.

A few years later when hundreds of navvies were excavating a new reservoir on the top of Hednesford Hills, Harry saw three of them pit themselves against three miners from Wimblebury in the old cockpit valley on the hills, and two wheelbarrows and a truck had to be used to remove the victims of knock-out blows.

To convince me that his abilities at 92 were still alert, Harry described how he regularly walked to Hednesford and back, doing the three miles, including a big hill each way, with perfect ease, and he challenged me to a bout of mental arithmetic.

"There's no chance of any of us getting a ha'penny change the better of him any time," was the daughter's supporting admission that Old Harry had all his wits about him. "And if he lends me a penny any time he likes me to give him one back which has the same date on it. That is one of his little whims at 92."

ON CANNOCK PARISH CHURCH TOWER

Men were engaged repairing the external stonework of Cannock St. Luke's Parish Church tower, and when I joined them on the roof it was after climbing the narrow, dark stone steps, and pushing open with the top of my head the heavy trap door.

My first observation was of the many initials cut or scratched in the big blocks of stones and on the chiselled pinnacles. Here and there were names which took the mind back to years of long ago, and several were reminiscent of similar sets of capital letters I had seen on the panels of the oak pews of the church gallery, which were dismantled in 1925. The contrast between the humdrum slates of an earlier restoration and the mellow diversity of the oldest portion of the roofing squares on the pyramidal centre slope was most noticeable. From the peak of the slates rose the anvil-hammered vane with its gilded head and tail.

Atmospheric clearness pushed out the sky-line to distant undulations where Salopian Wrekin and the companion lines of hills link hands with Clee and Clent, and to the lesser formations which peter out to Rowley. The beeches on Bushbury Hill looked like a fretwork screen against the silver sky. The fields on Cheslyn's domed Lodge Hill were harrowed red, preparing for coming green. Shoal Hill, the villages of Penk Vale, the scarped ridges about Castle Ring, a dozen pithead frames, the smoke of tileries along the Watling Street, a gleaming stream or two made the scene a varied one.

A group of slender attenuated spires revealed the presence of St. James's, Norton Canes, a church of fourteenth century foundation. On Leacroft rise were outlined uncomely remnants of the mansion of the Birch family: gateway pillars, dovecot and stables transformed into dwellings with fine oak stairs which are survivors from the sixteenth century.

A barge had just completed its climb up the fourteen locks from Churchbridge to Leacroft; on the Watling Street below motor lorries were hurrying with their mountainous loads between North and South, and overhead a plane was making of a hundred miles journey a task of less than that number of minutes.

On Calvan's Hill were familiar pines with a curious silhouette attitude of a galloping horse. Where Old Fallow mounted towards the open Chase were meadows farmed a century before by the stalwart Stubbs family who could furnish a team of twelve men who would be heavier than any dozen from any other family in Staffordshire.

Close at hand, on the south side of the church, was a business building formerly occupied by Crocketts, woolstaplers. In one of the store rooms David Taite Sheridan kept a school, and those who cared might see in that room a relic of the woolstapling and of the academy. It was an old oak wheel used in its turn for hauling bales from floor to floor, and for ringing the school bell.

Laughing disturbed Sheridan one morning when he was saying prayers. He threatened to thrash every scholar unless one of them could rhyme the cause of their laughter. The attack of the rod was held off when one of the lads stood up and said to the irate master :

> "There was a rat, for want of stairs,
> Ran down the rope to say his prayers."

Hard by was the redundant Rising Sun, stodgy in build, and stripped of all the glamour it had when it was a kind of side-chapel to the ancient church. It had a barrel organ with a very limited repertoire of tunes which was carried across the road to St. Luke's each Sunday morning for the services, and duly taken back to its rightful owner after evensong to provide more music in the Sun bar.

On the south-east side of the churchyard stood Cromwell House, whose new frontage hid the older Roundabout House built on an island site in the seventeenth century. It was owned in the early eighteenth century by Dorothy Birch, who made a bequest for the purchase of twelve penny loaves for twelve poor people of the parish on the first Sunday of every month. Dr. John Cronwell Blackford named the house after himself, and by mistaken corruption the name has passed from Cronwell to Cromwell House. No, Cromwell was never in the house.

When the eighteenth century was in its springtime residences of country gentlemen, academies, inns, shops of apothecaries, mercers, grocers and the like fringed the spacious market-place, and these could be envisaged in the intermingling low-built premises with dormer windows and the later loftier erections which effaced half-timbered originals. A welcome survival amid the modern transformations was the butcher's shop standing next to the Crown Hotel. It remained much the same as when it had a black and white farmhouse as its close companion, and William Drury combined the cultivation of a hundred acres of land with drawing prime home-brewed beer from the wood. The jasmine still made its golden festoons on the walls of the butcher's shop, in front of which were green lawns and flower beds with a simple fence.

Beneath the limes near to the preaching cross, was a colony of stones marking where men and women of Cannock Wood were laid to rest after they had been carried shoulder-high over rough tracks. Right in front of me was a collapsed table-tomb which was linked with a thrill memory. A miner going to work early one dark morning heard a rustling and the question "What time is it mate?" come from within the tomb, and he ran off as fast as his legs would carry him.

In the churchyard below me, five hundred years before, one William Griffyn, of Cannock, was humiliated by being compelled to walk, clad only in his shirt, in procession round the church on three successive days because of a serious moral infringement proved against him. At each circuit he had to bow down and kiss the steps of the church, and as he did so the priest castigated him with a rod.

THE BLUEBELLS RETURN TO LONGDON WOODS

There was a time when the bluebells of Longdon Woods provided one of the loveliest of Cannock Chase spectacles in springtime, and when the miracle of colour was luxuriantly wrought on the supremely beautiful ridge of larches, oaks and beeches which runs along the northern edge of Beaudesert Park and looks down on the fertile valley of the Trent.

Unbroken battalions of bells of deep blue hue spread as far as the eye could reach, and there was a delicate haze of colour rising to meet the fresh green of the new shoots overhead. There was a luscious peat to give strength and form to the bells, and the ground was littered with innumerable fibre needles. Nowhere else on Cannock Chase could there be found more flourishing wild hyacinths.

On many a day I saw the harvesting of bluebells proceeding, young folk from the neighbouring mining villages going out on foot and on cycles to gather loads of blooms.

Then came the great devastation of the tree-fellers. The ground was churned and rutted into a sorry morass; piles of debris were made into big fires which scorched wide areas, and when springtime arrived once more the blue carpet was missing, and no beauty could be seen taking the place of ashes.

These were the memories which came back to me one spring day in 1935 when the news was brought to me that the bluebells of Longdon had come back again. It was more welcome than tidings of the first cuckoo or swallow, or that the giant hawthorns, in Beaudesert Park were more profuse than ever in their cumulus heaps of white blossom. I was soon making tracks for Longdon, *via* Moor's Gorse and Wandon Lodge, and after I had made the descent and ascent of one of the most switchback roads on Cannock Chase, I came to where the bells held their carillon heads in the air.

They had won through in spite of merciless flagellations, and they seemed to have a more triumphant mien than ever. Yes, the haze of blue was once more lying on the hill-side and nestling like a mist in the hollows where slender streams make their way to the plain below.

Unpretentious striplings of trees were striving to flourish on rims of trenches scored by the reafforesters, resurgent bracken shoots were beginning to uncurl their fisted fronds, and the bunchberries, with their delicate green leaves shading into a gentle flush, were also proving that they were ready to renew their crimson carpeting.

The new Longdon Woods are as yet in their infancy, and it will be many years before they reach the stately growth of the plantations which preceded them. Another generation of gatherers of bluebells will have made their appearance by the time the larches have reared their ranks of protective arches of dark green above the belfry-floor.

PATSY THE KING OF CANNOCK CHASE

Patsy Kilgariff, the "King of Cannock Chase," had come to the end of his days, and I went through a torrent of rain to Chasetown to see him and his famous silken banner. It was his brother John who had asked me to make the visit, and when I knocked at the door of the cottage near the Miners' Rest there came the response, "Come inside, my love."

"Here is the king, sleeping his last sleep," said John, the elder brother, and before I had time to realise where I was the old warrior was being shown to me. He was lying in the little parlour with drawn blinds, and candles which were about him cast their flickering lights on the low ceiling and revealed curious decorations on the walls.

"Now love, we will talk about the king," said John, and many were the incidents and stories he had to relate. I asked to see the banner, and John's answer was "O, love, it is at his home yonder. You see he died here where he came when he was taken ill, but he lived yonder there in Paviour's Row by himself, and we will go, my love, to see the banner."

At the cottage on the way to Norton Pool dam, the blinds were pulled down to the bottom of the window panes, and when John turned the key in the front door he did so as if he were afraid of awaking the haunting spirit of Patsy. As soon as we entered the room there could be seen unmistakable evidence that a man and not a woman had been in charge there. The place was clean but in disarray, and no anxiety had been exercised towards the matching or toning of one thing with another.

Patsy's collection of portraits had been one of his favourite possessions. Most of the pictures were those of political champions whom he had been proud to follow, and there was also the face of Richard Croft Chawner, of Wall, into whose employ Patsy went when he came over from Ireland as a lad. There were numerous books about the rooms, including an inevitable life of the "G.O.M.," half-a-dozen entirely new volumes of a History of Ireland, and a handsomely bound edition of "Lives of the Saints."

Patsy's beloved Dictionary was there, and in it was a loose piece of paper on which were noted the meanings of orifice, orison and orgasm.

The banner was a splendid affair. It was made of red, white and blue silk, Patsy's adored red being by far the most predominant. The banner was about five feet square and was fixed to a gilt-headed staff about ten feet long. Inscribed on the banner in large capitals were the names :

GLADSTONE
SWINBURNE
FULFORD
WARNER
KILGARIFF

This was the banner which had been carried by Patsy through many election fights, and was fondly cherished by the "King of the Chase," a name expressing the leadership given him by his miner admirers.

When the redistribution of seats took place in 1880 and Sir John Swinburne was elected member for the Lichfield division, Patsy was tremendously demonstrative and the citizens of the cathedral city wondered what the world was coming to. They were terrified one day when they heard the "King's" men come shouting into their peaceful Market Place, and beating their home-made tom-toms with incredible unconcern in the sacred precincts of the Close.

Samuel, the Master Dictionary Maker, observing from his pedestal in the shadow of St. Mary's, might well have administered a mild rebuke to Patsy: "Sir, have you forgotten that word orgasm : immoderate excitement or action?"

The Mayor of the city demanded the arrest of Patsy," lest the huge Irishman from the Chase lays the city in ruins." Two hours of pandemonium, assisted by occasional re-fuelling at the inns, carried law and order to the verge of collapse, and then there was a sudden, impulsive right about face and the processionists swept round the turn by the George and trudged along the road towards Pipe Hill and back to the Chase.

The silken banner made its appearance one day in Shenstone, and as Patsy and his men marched through the streets of the village the "King" delivered fiery harangues and announced his readiness to tackle in argument the belted knights of the countryside.

Patsy considered himself to have reached the pinnacle of notoriety when in 1895 he had the audacity to heckle the great Joseph Chamberlain. The occasion was a big gathering in the waggon shed of Cannock Chase Colliery Co., at Chasetown, when Chamberlain spoke in support of the candidature of Major Leonard Darwin. Chamberlain invited Patsy on to the platform, and the progress of the "King" through the crowd was accompanied with tumultous cheers. Patsy argued, he repudiated, he denounced, he exhorted. He would not keep quiet, and he helped to make the meeting a life-long memory to all who were there.

One day I met Patsy walking from Lichfield towards Chasetown. It was at the time of an election, and he was wearing the Inverness cape and silk hat which he always donned for the campaigns. Patsy had lost his right arm by getting it entangled in a farm machine at Chawner's when a youth, and he took a fancy to the Inverness in order to cover his defect and to increase his resemblance to Michael Davitt, who also had only one arm and wore an Inverness. It was through the kindness of friends that he was found a position as a clerk at the local colliery, and given opportunities to develop his undoubted talents.

Patsy seldom returned from a stroll to Lichfield without bringing back with him a copy of Nuttall's, which he had bought for the purpose of giving it to one of his friends as a dictionary benefaction, with the characteristic advice: "Get this into your head."

He had a copy of Nuttall in his pocket when I met him, and he told me he did not think he was making an exaggerated claim in asserting that he knew the meaning of every word in the English language. He was often challenged by people to put him to the test, and when he answered with his etymological definition he would invariably illustrate it with a political allusion.

In his last days it was for Colonel Sir Courtenay Warner, M.P. that Patsy carried his silken banner, and never were two more contrasting temperaments happier in their common cause than were the "King of

Cannock Chase," the erstwhile farmer's boy, and the knight who was lord of the manor of Brettenham in Suffolk, and a son of Eton and Oxford.

CANNOCK CHASE, FROM RYE GREEN

THROUGH SPRING SLADE ON AN AUTUMN DAY

Two or three hundred yards beyond Pye Green corner in the direction of Broadhurst Green the road touches the highest point above sea level of any highway on Cannock Chase.

On reaching the brow of the hill from Pye Green it was apparent that visibility was far from good, and that the extreme skyline did not exceed two or three miles. There was a burning sun, but the atmosphere was occasionally toned down by a sprightly breeze. On the open moors the heather had lost its purple polish, and was bleached to a faint pink. The forlorn spikes of the faded red willow herb stood amid flaxlike festoons which were being shattered by the wind, and the flimsy tufts were carried away to tuck themselves in earth crevices till another year.

Along the Pottal Road there was no one afoot or awheel to be seen. Drawn up at the gate of the Military Cemetery were a couple of motor cars, and inside the laurelled fringe were several visitors standing round the white memorial on whose shaft the bronze cross had left its stains. Nearby, the old Cank Thorn had its autumn load of berries, whose scarlet and cream tints gave the venerable boundary tree a special dignity.

An effort had been made by someone to feed the large potholes in the metalled trackway with scraped-up rubbish, but the candid words of the finger-post writers still remained perfectly true—"Poor road." The driver of a large saloon car abandoned his effort to steer through the narrow spaces between the traffic excavations, and he and the friend with him set off on foot for the moors Sherbrooke way.

It was heavy footwork moving in and out along the churned-up trackway, which was in no better state than it was a couple of centuries before, when coaches from Birmingham to Stafford and Manchester made a straight course along the elevated ridge and emerged from the moors on to the Cannock and Stafford road at Brocton Gate.

Approaching Spring Slade Lodge the walking was more comfortable in the grassy grooves of the original track by the side of the plantation fence, this being only several feet from the later-day slag introduction of the military campers of the Great War years. While I was musing on the possibility of the dilapidated road being reconstructed by the powers that be for the benefit of visitors to the Chase, three women called me to a halt to ask the way to the military cemetery. They said they had hoped it was not much farther, as they had walked up from Brocton. They told me they wanted to see the grave of a

New Zealander whose mother had written to them asking them to buy some flowers and place them on her son's grave.

Instead of taking the turn facing Spring Slade Lodge which leads to Sherbrooke I went left down the track past the cottages. For a mile and a half the walk was through a continuous parallel of flourishing pines and firs. In sheltered corners there were a few banks of heather which retained their best purple. At intervals by the track side there were sturdy young oaks of from eight to ten feet in height. Some distant year they may reach the dimensions of the trees that gave the name of Oak Hill Plantation to now vanished woodlands thereabouts. The ubiquitous silver birches were obtruding their unasked-for presence in unplanned irregularity.

Isolated in a trough of a small gravel-pit was a bush that surprised, for every one of its hundreds of blackberries was as black and glossy as polished jet. I wish I had seen them before I met three small children who were trudging their two and a half miles home from school at Bednall.

All was delightfully peaceful and lovely in Spring Slade, and off towards Sally Harding's Valley the new afforestations held up their heads in undulating sweep. Large oaks and chestnuts were to be seen on nearing the Stafford Road end of the track. Piled at the sides of the narrow cart-road there could be detected beneath the thick undergrowth of years hundreds of tons of slag which, most probably, had lain there forgotten since military days when quick plans were made for the repair of Chase tracks, and as speedily pigeon-holed when the imagined emergency evaporated.

After a few minutes rest on the grass verge of the Stafford Road, a "Red" bus was boarded for Cannock, five miles away, Pottal Pool Lido being passed *en route*. It had taken two hours and a quarter to cover the route from Hednesford to Cannock, the distance covered on foot from Pye Green to the Stafford Road end of Spring Slade being four miles.

A GARDEN OF DELIGHT AT LONGDON GREEN

Descending from the high ridge of Upper Longdon through defile lanes, where vehicles can play hide-and-seek as they pass between rampart hedges and by casual field-gates, I came to one of Staffordshire's most spacious village greens.

On this Trent-side fringe of Beaudesert Park there was a lovely garden for which I was bound, and when I arrived there I found a welcome contrast from the noise and hurtling lorries of the great highway, which is never silent day or night.

In the garden was a clear, swift stream which had its source on the uplands of Cannock Chase. On the verge of the waters was a wonderful variety of plants, shrubs and trees. After passing under comely bridges of Japanese design the stream opened out into a charming pool, in which African flamingoes indulged their ease, their bodies of snow-white, with under-wings in gold, poised for long spells on a solitary leg, pink and slender.

Stanley cranes from South Africa and Demoiselle cranes from India lingered on the wide lawns that framed the pool, and there were glorious backgrounds of herbaceous borders whose colours rose in imposing tiers.

When Captain H. S. Stokes, M.C., was out in the Ypres salient he whiled away spells in the trenches by sketching a lay-out of these borders, and it was from the ideas he then set down that this wealth of beauty was brought into existence by his father, who carefully adhered to his suggestions as to plants and colours.

In after-war years the herbaceous borders had grown more lovely, while retaining their original tone effects. When Captain Stokes returned to Longdon after service in the Welsh Guards, he brought with him as a souvenir a bush rose he had rescued from the garden of a smashed-up cottage in No-Man's-Land, near Ypres. It was said to be one of "The Seven Sisters." Planted in one of the borders at Longdon, it had flourished vigorously, producing a mass of white roses touched with pink.

It was a garden of delight that Captain Stokes had created around him at the Brooklands. There was a choice blend of well-known and much-loved old English flowers and shrubs, with many striking ones from far corners of the world.

There was a carpet of autumn crocuses of white, gold, and pale blue. Clematis was everywhere in several various colours, which made arches, gateways, and even potting sheds decorative. There were striking vistas down avenues of loveliness. The stream, waterfalls, stately foreign ferns, tall rushes tipped with dusky pokers all in a row, bamboos, pink and white meadow-sweets, and gaily-plumaged birds all contributed to the colour scheme.

Oriental water fowl and ducks glided up and down the stream and made arrows across the lily pond. There were Bernicle geese from North Africa, maned geese from Australia, teals from Eastern Africa, Chiloe widgeons from South Africa, Bahama pintails from the Falkland Isles, and red-crested pochara.

As I strolled round the gardens I saw at liberty gorgeously-plumed Reeve's pheasants, golden pheasants from China, Impeyan pheasants from the Himalayas, black-necked crown cranes, Chukor partridges and Ruddy sheldrakes from India.

In the extensive aviary I saw a tournace which was the first to be reared in Great Britain. A flock of budgerigars provided in luxuriance all the colours of the rainbow. and the pink-cheeked, orange-cheeked, gordon blue and red-breasted wax-bills staged exquisite spectrum studies in thimblefuls.

A sundial set in the midst of soft-toned flowers commemorated that "H.S.S." won the Military Cross in France on September 25th, 1915, and in the shadow of a tree was a stone inscribed, "Herbert, 1917 to 1920," this being the Captain's memorial to a favourite tucan.

In this old-world garden, sheltered in a fold of typical Cannock Chase scenery, and enriched by an Empire's contributions of tree, plant, flower and bird, I loitered for several hours. When I returned once more to the village green, I passed from the fascinating to the desecrated. Cluttering the central greensward was a medley of paraphernalia parked by the tar-sprayers of the county roads. Abominably messy and hideously gawky, the implements were a despoiling blot on the village landscape, and I wondered if there were a county bye-law to deal with such acts of disorderly behaviour by the highwaymen.

Other thoughts came into my mind as I opened a wicket gate on the other side of the green and went inside a small, unpretentious building which I knew to be the oldest Congregational Church in Staffordshire. As I sat on one of the hard, bare benches, I fancied I could see in the pulpit some of the brilliant men who had charge of the conventicle long ago. There came to Longdon at the age of twenty-three eloquent Samuel Stubbs, son of John Stubbs, of Wolverhampton. A student of vast erudition and extensive learning, he consumed himself as a fire, and died at thirty-eight leaving his valuable collection of books to Manchester College.

Another Longdon preacher was John

Alexander, who was a friend of Dr. Joseph Priestley, and was accounted to be one of the best Greek scholars of his time. One Sunday morning when he was due to preach at Longdon he was found dead in bed.

William Salt, born at Cannock in 1783, usually preached at Longdon on Sunday afternoons. On one occasion at Lichfield there was a man in his congregation who, on the following day, committed suicide. The jury returned a verdict that the young man's deed of self-destruction was the result of the effect upon him of the terrifying doctrines he had heard preached at the Dissenting Chapel. This finding provoked a tumult in Lichfield and Longdon, and in order that the world at large could judge for themselves whether his sermon was calculated to drive people to suicide, Salt delivered what came to be known as "The Killing Sermon" up and down the country. He turned these special occasions to practical use by devoting the proceeds of the collections to the building of a new chapel at Lichfield.

A POETESS AND THE BESOM CLUB

Shareshill was full of people when the annual parade and feast of the Besom Club took place. At the head of the procession was a man carrying a mop stale with a large bunch of flowers decorating the top. Halts were made at all the public houses, and the excesses of conduct and speech were not less than had been anticipated.

There was a certain little girl who had a natural desire to see the fun, but she was solicitiously whisked away by the vicar's daughter, who was fearful about the ill-effects of the Besom Club incidents on the young folk of the parish.

The girl was Eliza Wootton—later Mrs. Eliza Pearson, of Restdale, Cheslyn Hay—and the vicar's daughter was Frances Ridley Havergal, whose fame as a hymn-writer has gone throughout the world.

It is not surprising that Mrs. Pearson enjoyed for many years a reflected glory because of her association with Miss Havergal, and when I called to see her one afternoon she had much to tell me about Miss Fanny.

"She used to take me into the dining-room and give me most fascinating Bible lessons. I have never heard anyone who could describe things in more beautiful language, and, what is better still, I have never known anyone who was more beautiful in manner and more kindly in disposition," said Mrs. Pearson.

Mrs. Pearson spoke of the poems written at Shareshill vicarage by Miss Havergal, and of the talent she had—like her father—of musical composition. The vicar gave many of his anthems and hymn tunes trial runs at Shareshill Church.

As a diversion from the excitements of the Besom annual the villagers set up an agitation against a proposal to institute a Sunday delivery of letters in the parish, and Miss Fanny was most vehement in her denunciation of a project which would bring about a wicked desecration of the Sabbath.

One day two nieces of Miss Fanny and Eliza fell into disgrace when they were tempted to go to a private room and look at a lovely new blue frock which was hidden beneath masses of white tissue paper.

Miss Marie Virginia Georgina, sister of Miss Fanny, surprised the culprits, and she terrified them by saying in grave tones: "You know how it would be if you put your finger in the red-hot fire. Well, that is how you will be horribly burned for ever and ever if you are disobedient."

When Eliza left the vicarage for Cheslyn Hay, the vicar gave her a Bible, and, placing his hand on her head, said: "I pray you may be kept from the evils of the world."

That was just how Miss Fanny felt on the Besom Club day about little Eliza, who, at eighty years of age, was full of pleasant memories of the vicar-musician and the poetess of Shareshill.

"They were both very fond of Cannock Chase beauty and often talked about it," said Mrs. Pearson, "but they always declared that the loveliest thing in Shareshill and all round the Chase was goodness."

"Of course that is old-fashioned talk, I know, but it is no more out-of-date than the pretty places of the Chase of which you are so fond of telling us," she added.

COME AND SEE CANNOCK CHASE

"Come and see Cannock Chase" was one of the warmest invitations which Edward John Littleton, the first Baron Hatherton, was constantly extending to his friends in all parts of England. There were generous hospitality and many pleasing things to see at Teddesley, and no visit was complete unless the guests were taken over the hills on to the open moors and into the valley of Sherbrooke.

It did not matter which season of the year it was, the Chase had a mood with which it was worth making an acquaintance. Judge J. N. Talfourd, poet and biographer of Charles Lamb the whimsical, stayed with Hatherton when on Assizes duty at Stafford, and they had many a fine tramp on Cannock Chase. They had planned another stroll one day when Talfourd died suddenly in the Stafford Assize Court as he was in the act of delivering judgment.

Anna Jameson, who was a noted art critic before John Ruskin, began her literary career while she was a governess at Teddesley, and in later years she was often on Cannock Chase with Hatherton. One day she told her host as they went over the moors how she had befriended Robert Browning and the young wife with whom he had eloped, on their arrival in Italy.

Mary Russell Mitford, author of "Our Village," saw the Chase with Hatherton, and related to him how as a child of ten years of age she had won £20,000 in a lottery, her success being luckily due to the fact that she insisted on choosing a ticket with the number 2224 because its digits made up the sum of her age.

Dr. Buckland, the geologist Dean of Westminster, had a rare walk over the Chase, and caused amusement when he produced from his pockets curious fossil bones just at the moment when some particularly attractive Cannock Chase scene came into view.

John Peter Labouchere, one of the founders of Deacon's Bank, and father of a more noted son, Henry Labouchere, editor of "Truth," had an exciting time when he set out over the Chase, and had to negotiate huge chasms which had been made in the tracks by great falls of rain.

Lord Redesdale, Speaker of the House of Commons, had exceptional delight as he went into some of the loveliest parts of the wild Chase. Lord Lyndhurst, a feeble man of seventy-eight, rode over the Chase on a September day in 1849, and as he went by Pottal, Boss Green and Penkridge Bank, the great politician mingled political talk with exclamations of delight with the scenery around. At night his host read Shakespeare to him.

Charles Greville, the famous diarist, whose indulgences in private details of the great shocked Queen Victoria, was on Cannock Chase in September, 1843. He found much pleasure in his rambles, but those who entertained him were astonished at his wearing a flat tenpenny labourer's hat in which no respectable person ever thought of appearing on the moors. He was ill-dressed, had no neck-cloth and was dirty, and was not thought to do credit to the scenery of the Chase.

"I love this kind of country—the lovely Chase," said the Iron Duke when he rode from Cank to see his old friend, the "Waterloo" Marquis of Anglesey, at Beaudesert.

AN UNSOLVED CROSS KEYS MYSTERY

The Cross Keys, Hednesford, has an unsolved mystery: did a big burly navvy who was doped there recover and clear away as best he could, or did he succumb and the notorious Palmer the poisoner secretly dispose of the body?

The incidents occurred when Bob Brettle, the noted Birmingham prize-fighter, was training at the Cross Keys for his great championship contest with Jem Mace. Brettle's principal backers were Dr. William Palmer, of Rugeley, and Cook, the Doctor's young friend and ultimate victim, and they made a practice of meeting their ring fancy every day at the Hednesford hostelry.

A roystering bully amused himself by playing all sorts of monkey-tricks in the taproom, one of his galling pranks being to pick up the glasses and cups belonging to the company and drink the contents with a blatant insolence.

Brettle saw the navvy's performances, and his blood got to boiling heat. "I can't stand it any longer, and I shall soon slip into this cad," said Bob.

"Don't be a fool; don't risk getting hurt for such a brute. Leave him to me. You go to the tap and get a quart of ale, and ask him to have the first drink. He won't refuse,'' such was the advice given by Dr. Palmer to the boxer.

As Bob was carrying the ale along a passage Palmer intercepted him, and, taking a small paper parcel from his waistcoat pocket, he shook a fine powder over the surface of the ale.

The navvy almost cleared the quart away with his first pull, and he left no dregs in the glass. For some little time the bully's voice was noisier than ever, and then he suddenly collapsed into a heap on the settle in the tap-room.

At closing time the landlord tried in vain to shake the sleeper into consciousness. Brettle lent a hand, but he had no more success, and so a bed of straw was made in the stables, and to it the helpless navvy was carried.

Bob had a bad night, for he had dreadful suspicions of what might have happened as a result of the navvy having drunk the ale which he had supplied to him. His first act the next morning was to rush to the stables to see if the dope had worked off and the navvy had vanished.

No, the bully still looked as dead as a door nail. At the end of twenty-four hours Bob was so convinced that the navvy was beyond recovery that he felt inclined to go to the police and explain the situation. He was afraid, however, that such a course might implicate his friend Palmer, and so he resolved to bide his time.

Bob paid many more visits to the stables during the day, and before he retired for the night he felt he could not resist going again, by the light of a candle, to see how the navvy was faring.

The man had gone. But how, or when, or dead or alive? There was not the least evidence or explanation.

Bob could not make up his mind whether the devil or Palmer had stolen the body, but, later, when some of the Doctor's fiendish deeds came to light, he awarded the credit to the man who sprinkled the powder in the glass of ale.

Bob also averred it was this episode in the life of Palmer which first induced the suspicion of his being a poisoner.

JACK OF HILTON EMERGES FROM HIS GREEN BAG

"Jack of Hilton" is the most notable character in Staffordshire antiquarian lore associated with New Year's Day. He is one of the most curious and mysterious images in all England, and it was in order to make acquaintance with him that I went to Hilton Park.

No one can speak definitely of his age, his origin, or of the significance of the extraordinary custom connected with him for centuries. The fact is certain that eight hundred years ago "Jack" performed his special function of blowing the fire every New Year's Day in the great kitchen of the ancestral home of the Vernons at Hilton.

For about two hundred years he acted his part in the feudal ceremony between the lords of the manors of Essington and Hilton, and then he was pensioned off from active service and placed on the family sideboard handy for use on festive occasions.

For five and half centuries "Jack" has been carefully preserved by the Vernons, but in more modest days, because of his crude nakedness, he has but rarely come into view of visitors to the hall. A century ago Henry Charles Edward Vernon Graham thought fit to relegate "Jack" to a green canvas bag, neatly tied with a piece of red tape, and duly docketed with a card on which were placed the words: "This brass image belongs to Colonel Graham."

When I went to Hilton Park to look at "Jack" it was not with hopes of seeing him fill his old-time role by the side of a roaring fire, for the days of the immense kitchen blazes are no more seen in a mansion where central heating is now firmly established.

Mr. Walter B. Vernon, son of the noted "Squire A.L.V." of many local associations, is now the guardian of "Jack of Hilton," and he it was who untied the green bag and brought the ancient fire image into the light of day for a while. As the owner placed "Jack" on the study table he said, "I hope I shan't have bad luck again after this deed, for the last time I fetched him out of the bag I had the misfortune to meet with two accidents."

A glimpse at the image certainly provoked the thought that the unlovely-looking fellow might easily have a lot of devilish mischief in his make-up. We set him on the table, and at once began to comment on his heathen-idol cast of countenance, his thick-set body that conjured up notions of evil might, and we wondered what mediæval craftsman had designed "Jack's" suggestiveness of general pose.

"Jack" is twelve inches in height, is made of bronze and weighs eight pounds fourteen ounces. He is stooping, with his left knee on the ground, and his right leg is pushed forward so that the lower part of the trunk is hidden to the onlooker from that side. His right hand is placed on the top of the head, and the left hand rests at the front of the body. He is naked and unashamed.

There is a small hole in the middle of the back, and a still smaller hole in the mouth, the latter aperture being capable of admitting only a thin wire.

Water was poured into the hollow figure through the hole in the back, this inlet being afterwards stopped up. "Jack" was then set by the side of a strong fire which caused the water within to evaporate through the mouth of the image, "with the same manner as in a aeolipile," to quote the words of one who saw "Jack" give a demonstration for his benefit two hundred and fifty years ago.

The blast that came from the mouth was so considerable that it made a loud noise, and blew the fire with such force that it was made to blaze furiously.

The occasion of "Jack of Hilton's" exhibition of his skill was when the lord of Essington had to take—in accordance with the conditions of the tenure of his lands—a goose each New Year's Day to the lord of Hilton and drive it round the kitchen of the hall at least three times while the image blast was blowing. The goose-step concluded, my lord of Essington handed over his gift to the head of the house of Hilton, and the latter, with due ceremony, passed the goose to the head cook, who received it with the obeisances that raised the kitchen for a while to the status of mediæval chivalry.

Trussed and set on the spit, the goose was soon colouring to a succulent russet; "Jack" went on making a fuss with his puffing and blowing, and the manor lords clicked their tankards and rattled their tongues about the latest adversity of agriculture.

"Ready, my lord," calls the master of the roasting, and he gives back the goose to Swynnerton of Hilton, who hastens to slice a fair portion off the breast, and gives it to the lord of Essington for his meal.

There was fitting courtesy and gracious formality in all this, and it is not difficult to see in the tribute of the goose an edible variant of the pair of white gloves, a clove of gilly-flower, or a pair of gilt spurs, which were so often tenure payments in those days.

The introduction of the bronze image remains a mystery. Most likely "Jack" was brought on the scene and attached to the goose ceremony by someone with a taste for the unconventional and a desire to make use of a novel figure which had come into his possession. Antiquaries have offered differing views about "Jack." Some think he is of Etruscan origin, and was brought to England in Roman times. Others speculate that he was originally an Oriental idol who was worshipped as a representation of the source of human life, and was brought to England during the Crusades.

It was when the Swynnertons became lords of the manor of Essington as well as of Hilton that the goose and image ceased to play their concerted tenure parts in the baronial kitchen. That was in 1348, and he who carved the last tribute goose was John de Swynnerton, who was Sheriff of Staffordshire in 1361, and died in 1379. For two and a half centuries the Swynnertons held the Stewardship of Cannock Chase.

"Jack" continued to display his talents to visitors to Hilton Hall long after the supply of tenure geese ceased. When Robert Plot, the county historian, went to Hilton on May 26th, 1680, he personally came into contact with men who had been present fifty years before at a replica "Jack" and goose ceremony in the hall kitchen.

Others besides Plot were favoured at various periods with tenureless revivals, and the present owner of "Jack" can recall days of special jollification at which the family hero did a special turn.

My parting acquaintance with "Jack" was to lift him and look at him in front of a westward window through which the departing sun was sending waves of golden glow. Here was "Jack" in a new setting. Once again he was athwart a fire, but the sun did not stir him. To both of them, in divers times, men had fallen down and worshipped.

As I handled "Jack" I heard something moving about inside him. Curiosity could not be satisfied, for no eye can pierce therein, and in the twentieth century, as in bygone hundreds of years, "Jack" will retain his secrets.

THE DEERS' DEAD HEAT DOWN WOODLAND TRACK

When I set out on my three hours saunter of the Chase one November afternoon, there was a blanket of mist hiding everything beyond a mile radius, and the sun was visible as through a smoked glass.

The distorted silver birches which lined the roadside from Pye Green to Fisher's Lodge were bereft of the tawny leaves that a few days before seemed loth to let go their hold. On the opposite side of the moorland highway the hedgerow was matted with moist and mellowed bracken. Boss Green had thousands of added lengths of plumes crowning the tops of new forest trees. Down Pottal Road, now smoother for feet and tyres than at any previous time, faintly swaying woodlands presented a vista which had no relief of colour of sky or tree at its perspective climax.

Down a narrow gully through close ranks of flourishing children of the trees I made my way, and, striding over a broad track, entered a mass of forty to fifty years old pines and firs, where slashes of grey sky lay on the tops of the branches. The coating of needles made it appear as if the ground had been strewn with sawdust. There was no undergrowth to encounter, for here in the lower realms there was no sunshine life, and not a solitary bird was there to give a greeting. I wondered if that was one of the penalties of a Chase buried in what some of the cynics had described as Christmas Trees.

The trees rose to a height of sixty to seventy feet, some of them in evergreenness and others in winter shabbiness. In the heart of the wood it was like walking among a host of dingy flag-poles whose pennants were out of sight. No wonder the place bore the name of Dark Slade Wood. I came to a hollow where it was "Crackle, Crackle, all the way," for the fallen decayed lower branches of the trees had been accumulated in a great mass to a depth of two feet. It was easier to face the obstacle than to retrace my footsteps. I thought if there should be deer about that afternoon I was doing sufficient snapping of twigs to send them scampering.

Emerging from the underworld I went over a trackway and dipped into another plantation where the growths were about ten feet in height. So closely had the planters set their saplings that passage was now gained only by twisting and stooping, curving and sideway thrusts.

Heather with pink nodules on upper tips kept warm the feet of youthful larches, and in shallow dips between stubby bosses of benty grass there were casual bilberry wires to which vagrant leaves of ruddiest hue were clinging.

In this thicket I wandered for some time, and then once again I found a broad clearing, about one hundred feet in width and half-a-mile in length. As I stepped into the open a couple of deer were standing perfectly still in the middle of the avenue. They must have heard me approaching.

They were young and beautiful, and they proved they had the agility and dash of dawn-fresh days. As soon as they caught sight of me they shot away with magnificent ease. I clapped my hands just once with instinctive surprise, but there was no need of such stimulus to speed. Right down the track they bore, side by side, about three feet apart, never once halting or diverting to the right or to the left.

At the end of the straight half-mile the deer plunged into another sister shelter of the Dark Slade. It was a dead-heat at the end.

Never have I seen four-footed creatures race the course more swiftly than those startled fawns about the Warren Hill, Badger Slade, and the woods of Dark Slade.

When I returned to the road again the sun was enjoying a little more mastery of the mist. There was a ribbon of gold flung athwart Pottal Road, and a cotter was bringing his pony from croft to stable for the night.

ON THE BAGOT LANDS FOR SIX CENTURIES

Tree-fellers were busy among the Bagot oaks, and the crackle, crunch and thud of collapsing centenarians could be heard all day long.

A fleet of motor drags was hauling the timber to Grinley station, six miles distant, where three Bagot oak "specials" steamed away every day. Exits from the woods were churned into slushy, mud craters, out of which teams of horses had to pull the despairing eight-wheeled petrol lorries.

Acorns lay about in prodigal profusion, and colonies of pigs feasted and fattened without herdsman's interference. Thousands of Bagot oaks were disappearing, but in the thinning out process thousands of comparatively youthful ones were being permitted a longer lease of life.

The age of Elizabeth, of the Stuarts, of the Georges were being cleared out of the Bagot woodlands, and more daylight would be available for the oaks of the Age of Victoria. Much of the timber which was being felled was found to have gone a century past its prime, so loth had the lords of Blithefield of recent generations been to bid farewell to their historic woodlands.

Home-breaking was also proceeding in the

Bagot countryside, and as I walked in woods and lanes that afternoon I learned of two sisters who were anxiously looking out for a cottage for their eventide. Miss M. E. Brown and Miss E. Brown lived at the Nursery Farm, Heatley, not far from the mansion of the Bagots, and they claimed that their family had been on the Bagot estate for six hundred years. Their ancestors came from Wales to Staffordshire with members of the Bagot family, who had estates in the Principality. Each generation the Brown succession had been maintained from father to son for six centuries. Ill-health caused the latest male successor to seek other occupation, and for thirty years the two sisters had carried on themselves at Heatley.

They gave me a cordial welcome to their farmhouse, and told me it was so seldom they came into touch with anyone from the outside world that they felt excited to have a visitor. I felt their elation had a descent when they found out that I had not come to tell them of a house that might do for them.

They could offer no documentary evidence of their family's six hundred years' occupation of Bagot lands. It had been a tradition handed from father to son for centuries.

They had an old chest, a massive Family Bible with "s's" like "f's" and a collection of Toby jugs and other links with the past, but no deeds, leases, bills, letters or anything that revealed authentic antiquity.

They understood one of their people one day many years before, burned a great quantity of papers which were said to be no good to anybody, and they had often wondered if the fire consumed records that would have told the story of the Browns of the Bagots. A few cabinet and *carte de visite* photographs which always provoked smiles, had been spared from the flames.

Were they lonely in that out-of-the-way corner of the world? Of course they were not, for even though they had the dark winter they had the country delights in the summer. In addition, they now had the wireless.

The apparatus occupied a good part of half of the kitchen table. It was an old-time crystal set, and earphones were lying ready for use. The sister Misses confessed they would dance for joy if a message could come through telling them of a house that was waiting for them. I promised I would keep my ears open in places were spare houses were sometimes talked of. When I met the village policeman I sounded him on cottages, and had a surprise when he informed me that his sergeant had bought the Nursery Farm at the sale and was going to retire and live there.

The next person I met on the road soon forced me to the conclusion that it is a habit to think in centuries down Blithefield and Abbots Bromley way. He was a roadman engaged in clearing away the thick mud left by the timber motors on the highway. He was proud of the fact that his ancestors came over with the Conqueror and an early Bagot. Carpenters, sawyers, game-keepers had been the men of his line, and they had always worked on the Bagot lands.

He boasted that his grandfather brewed the beer which was drunk at Blithefield when the late Lord Bagot was born, and a supply of which was kept to be consumed when his lordship reached the age of twenty-one.

"Something like beer it was in them days, as my father and grandfather used to tell me," said the roadman.

When I walked into Abbots Bromley everything in the village seemed as quiet as the old dobby horse resting in the transept of the parish church. The noted St. Anne's and St. Mary's Schools were closed for the holidays, and the many hostels in the main street had closed doors and windows.

The seventeenth century market cross or butter market presented a remarkably improved appearance after its restoration under the supervision of the Society for the Preservation of Ancient Monuments, at a cost of £200. The roof slates had been replaced by antique tiles and strong lead ribs had been introduced at the hexagonal lines. All the timber-work had been repaired and strengthened, and the former earth floor had been superseded by a diamond-shaped surface of cobblestones.

Over seven hundred years ago Henry III. made a grant to the monks of Burton to hold a weekly market at Abbots Bromley. It was here on the site of the restored cross that the market was held, and it was when Henry VIII., at the Dissolution of the monasteries, pocketed the possessions of the Abbots of Burton, that Abbots Bromley cross was given by him to the Paget family, which had now handed it over to the local parish council to be protected and preserved.

WHEN JOHN ORDERED OUT THE DUCHESS

It was John William Bailey's unusual distinction to have been a boy solo soprano at Birmingham Musical Festival in the "Messiah," "Judas Maccabeus," and Mozart's "Twelfth Mass;" to have been the third lad to go down the Cannock Wood pit to work, and to have been chef to a duke, a marquis, and a duchess.

John was getting on for ninety years of age when I met him in unexpected circumstances. One Saturday afternoon I was looking round the garden of the seven hundred years old St. John's Hospital, whose row of curious, gaunt chimneys are a striking feature of Lichfield's main street, when I got into conversation with one of the dozen aged men for whom a peaceful and comfortable shelter is found.

It was John Bailey, and he asked me if I would like to have a peep at his room. When he found that I came from Cannock Chase he became quite excited, for he said he knew all about the Chase, and could tell me a great deal about people he had known there many years before.

John was greatly interested when I told him that hundreds of years ago by a grant of one of the Bishops of Lichfield, two loads of wood from the forest of Cannock Chase were granted annually to each inmate of the hospital.

John's mother rang the Lichfield Cathedral bell for the services daily for sixty years, and, as a lad, John gave her a hand now and again. When he became a cathedral chorister he had to bid adieu to the bell. In those days George Augustus Selwyn was Bishop of Lichfield, and his lordship was greatly pleased when the bell-ringer's lad was honoured by being chosen to sing at Birmingham Festival along with famous vocalists of the day in the great works of Handel and Mozart.

"I was made a fuss of and was given plenty of presents because of the way I piped out the high notes at the Town Hall, and I was very sorry when my voice broke and I couldn't take the soprano parts any more," said the old man.

It was not long before he was sent to London, through the kind interest of Lichfield people, to be apprenticed as a chef at a large establishment near to the noted "Dirty Dick's." He found the hours too many for his liking, and so one Sunday morning when the master was at church, John and two other apprentices ran away.

Instead of returning to Lichfield, John thought he would choose somewhere not quite so sleepy. He had heard of the good money to be earned in the pits of Cannock Chase, to which men were flocking from many miles around, and so he tramped to the Cannock Wood pit, where he was put to wheel coal from the stalls in a low-built barrow. He had changed from white to black at one jump, and the contrast amused him, but he was able to have his liberty after the day's hard work.

The miners soon learned of the youth's vocal talents, and he was in frequent demand at the Cross Keys, the Trafalgar and the Anglesey for "free and easies." When Hednesford Church was opened in the 1860's the ex-cathedral chorister sang in the choir, and nearly seventy years afterwards he was able to reel off a long list of names of men and lads who led the singing that day.

He soon got tired of the pits, and decided to take to his chef work once again. He was fortunate to get the backing of a recommendation from Bishop Selwyn, and he went into the service of the Duchess of Sutherland at Trentham Hall and Dunrobin Castle.

John burst into laughter when he mentioned the Duchess, and, slapping his hands on his knees, told me of "a lovely row" he had with her grace. He had prepared what he felt sure would be a dainty morsel which would please everybody, but it did not please the Duchess.

She went into the kitchen in a haughty

manner and scolded John, her cross words not being as polite as John thought they should have been.

"So I just showed her to the door and ordered her out, and told her as she went off in a terrible temper that she had better stick to her own job and I would do mine," said John. "But you can guess what happened after that. I had to clear out, but as I had married the Duchess's wardrobe maid only a few days before I had something good to bring away with me, as I had reason to know in after years."

Nine years as head chef for the Duke of Portland at Welbeck Abbey and several years for the Marquis of Anglesey at Beaudesert Hall followed. That was the heyday period for John, and with money he had saved the chef decided to start in business as a grocer. Hednesford attracted him once more, and he opened a shop there in the main street. At the end of ten years he was moved by the roving passion again, and he bade Cannock Chase farewell.

After some years of wandering he found his way back to Lichfield. He was a poor man, and feeble with years, and was thankful beyond measure when the doors of St. John's almhouses opened to admit him.

It was in the room in which John lived, and slept, and which he was expected to keep clean and neat, that he gave me glimpses of a life which had never lacked incidents worth remembering and relating.

"Now don't try and remember what I am going to tell you now, for it's just one of those bits of devilment that we used to enjoy on the Chase sixty years ago," said John, and he followed his hint with amusing accounts of tricks played on folk who lived at Cannock Wood and Rawnsley.

The practical joke which seemed to tickle him most was played when a widower left his cottage empty while he went to Lichfield market. When he arrived home he found that a pig tied in a bag, with a blue poke bonnet on its head, had been placed in his bed.

Like all other occupants of St. John's the old man had to cook his own meals, and he was always ready to give chef wrinkles to his companions who were not so well versed in the arts of making simple morsels sumptuous.

THE LADY WITH THE DUSTER

Standing one day on a hill overlooking Brereton I entered into conversation with a man of seventy odd years, who, like myself, was loitering about enjoying the view. He talked to me of the various features of that part of Cannock Chase, and specially called my attention to a broad tower which rose above the woodland setting on the far skyline. It was Hoar Cross Church, and the stranger began to describe to me the occasions of years before when he used to walk with his father from Brereton to the church and back to hear the silver-tongued vicar, Canon Knox Little, preach. The Canon had the reputation of being able to fill St. Paul's Cathedral as easily as he could the Staffordshire Church of the Holy Angels.

On the Brereton hill-top that day I heard of the wonderful eloquence of Knox Little, and of the way in which his sermons thrilled the packed congregations of the village church by their power of persuasive felicity. First he would lash out in torrents of vehement denunciations, then he would soar to the heights of brilliant imagery, or sink into a low murmur, and finally end with dramatic abruptness.

Our talk passed on to the striking faithfulness of Viscount Halifax, the first chairman of the Trustees of Hoar Cross Church, a duty placed upon him by his sister, the Hon. Emily Charlotte Meynell Ingram, the founder of the church. For over fifty years he cared for the building with untiring solicitude, and there was no repair which was too small for him to supervise. He was often at Hoar Cross, and when the news arrived that he was about to pay a visit, everybody about the place had to see that all was spick and span for his inspection. In his early days at Hoar Cross he met a band of brilliant men who were destined to be in later years in the front rank of dignitaries of the church. Edward Talbot became Bishop of Winchester; Francis Paget, Bishop of Oxford; Charles Gore, Bishop of Birmingham; Henry Scott Holland, Canon of St. Paul's; and Illingworth, Lyttelton, Richmond, Cheyne and Locke were others who gathered and discussed vital beliefs which, in due course, in the pages of "Lux Mundi," shook to its foundations theological thought in the National Church.

On the south side of the church at Hoar Cross may be seen the Viscount's face in stone, the sculptor having given St. Athanasius the countenance of the Chairman of the Trustees. It was a countenance which had a particularly wistful cast when the Viscount sat in one of the spacious pews dreaming of days of long ago. The "Lady of the Duster" who saw his lordship in pensive attitude many times had not the least doubt he was thinking of his saintly sister who first conceived the first dim outlines of the commemorative pile; who watched its rising day by day with keen and loving interest, and often sat by the hour on a chair within the framework, and, later, within the finished walls, telling first one workman and then another how she desired the hundred and one details to be carried out.

She did not hesitate to give orders for a piece of work to be done over again if it failed to realise her expectations, and more than one mason found his work cast aside so that there might be a worthier expression of the ideas of the dreamer put in its place.

It was the "Lady with the Duster" who told me things which the stranger on Brereton hill could not possibly know about the Church of the Holy Angels. She has been at the heart of the shrine for many years. It is no mean office to be the "Lady of the Duster" in such a church. She is able to detect specks of dust where the ordinary human eyes see nothing but oak or stone. The duster is always in her hand; she never seems to put it down; it goes either under her arm or over her shoulder when she wishes to open a door or to perform some other menial duty.

She points with the duster, and as she speaks she gives the duster a flick and away goes some trespassing atom. She opens the exquisitely carved Stations of the Cross to show how cleverly the artists did their work, and before the doors are pushed back again the duster travels a few inches to and fro over the surface.

Here lie the surpassingly impressive alabaster monuments of the founder and her husband, and as the Lady of the Duster talks in superlatives of the characters of the two who shed the radiance of their saintly spirits on the village and the countryside about, the square of cloth is passed over the faces of the sleeping pair. From the dimples of chin and curves of neck, from the marvellously carved lace around the shoulders of the foundress and the realistic uniform of the erstwhile yeoman and Master of the Hounds, the duster disperses the accumulation of dust of the last hour or two.

Wherever the Lady of the Duster takes you—in Chantry, in Lady Chapel, in aisle, in pews—the duster is for ever on the move, and as the visitor loiters for a time to hear the caretaker-guide tell things both new and old of the lovely church, the cleansing hand continues its ceaseless movements. That duster has never been dedicated by priest or bishop, but so beneficent is the work it does, and so devoted is the heart that beats behind it that surely it has the blessing that comes from service well done.

"They tell me I shall have to have this placed on my coffin when I die," said the Lady of the Duster.

JIMMY TELLS HOW THE COAL WAS FOUND

When "Jimmy" Harris learned that an official repudiation had been made of his claim to have found certain coal seams at one of the local collieries as the result of five visions he had seen, he sent for me in order that he might tell the full story. "Jimmy" had worked at the colliery for over thirty years when the management decided to explore for further new seams, and he was one of the company men employed on the operations.

"They deny that my word is true, do they?" said Jimmy. "How can they dare to say such a thing? Are they afraid to give God the glory? They know very well that what I say is true, and I challenge them to deny it to my face, or to meet me in public and prove whether what I say is right or not. You know what happened to Ananias, don't you? Well, that will happen again if they tell such untruths in my presence."

"You want to know what the visions were and how I had them, I suppose," said Jimmy. "Well, it was like this. They found the coal at one place and then lost it again. The management said it was a bad egg, and that the search had better be given up. So I said to them, 'I will find the coal, and also lose it again, and then find it once more. All the seams are there, for the Lord has told me so.' "

"They were so struck by what I said that they told me to get on with the job, for they believed that if Jimmy said he would do it then it was certain to be done. The heading was continued and within four feet from where operations were started the coal was found again. It was also lost and found again according to the vision."

Jimmy declared that when he had his first vision there were no signs of the coal in the place where they were heading, but the management had been of the impression for twenty years that there might be coal there.

One of the miners working with Jimmy exclaimed, "Well, Jimmy, thy God's a wonderful God and nobody can deny Him."

When the old miner began to tell me about the visions I asked him what he meant by certain words he had spoken.

"Ah, my lad," he replied, "God puts things in allegory to-day like he did in the days of Peter and John. He never talks flashlike, but he puts the truth in a way that you don't understand unless you have the Eye of Faith."

Did Jimmy see God in his visions? Of course he did, so he assured me. "I saw Him stretch forth His hand and show how deep would be the seam of coal. The measurement

He gave me was five feet eight inches. I told them that at the colliery next day about the measurement and promised one of my mates that if I was wrong I would give him sixpence. I was only two inches out, and I gave the man the sixpence.''

Jimmy told me he had been having visions for many years, and had the five visions very close together. It was about the time of the West Stafford election when Harry McLaren and George Lloyd had a set-to for the Liberals and the Tories. The first vision was, "Seeing the coal measures;'' the second, "Seeing the seam above the other seam;'' the third, "Seeing the drops or faults to be met with in getting the coal;'' the fourth, "Seeing a new colliery without shafts, the seam being so extensive,'' and the fifth was "Seeing the fruit and blossoms hanging round the colliery offices.'' The visions came to him in his sleep.

Calling me into his little parlour, Jimmy showed me a portrait of a lady which occupied a prominent position in the room. News of an extraordinary miracle of healing worked by Jimmy had reached this lady in London, and she wrote to him asking him to pray for her so that she might have the use of the hand and foot which had been practically dead for fourteen years.

"I told her of the day she would get the use of her limbs,'' said Jimmy, "and on the very day I mentioned she was marvellously cured. She was full of joy, and she sent me that photo and a lovely and costly case of jewelry which I keep upstairs."

"It shall be half-gone next Monday and on Tuesday shall be gone altogether," was what Jimmy said to a young woman who had been ordered by doctors to go into hospital for an operation for a serious growth in the mouth.

When the doctor went to see her on the Wednesday he could find no trace of the tumour. This was one of several achievements enumerated to me by Jimmy.

Then there was a personal affair. On one occasion he had an arm broken in four places while working in the mine. He was told he would never be able to raise the arm again.

"That dunner rest with thee; if God wills that I shall have the use of my arm it shall be so," declared Jimmy.

"He wills that I should raise my arm," shouted out Jimmy, and in the doctor's presence he immediately put his arm in the air, and from that moment the limb was restored to full use.

Before I left him, Jimmy showed me a copy of a letter he sent to the manager of the colliery company about the visions and the seams: "I asked my Lord to show me where the coal was and He said He would show it me. I took Him at His word, and ventured all I call good and true in the world."

Jimmy was seventy-five and his wife seventy-three when in later years they celebrated their golden wedding. They had been married at Willenhall where they were both born. When Jimmy lived at Portobello he was known as "The Giant Red Hen Flyer," because of his specialities of pigeons. When he moved to Cannock Chase he came under the influence of the Methodists, and his religious fervours became as striking as had been his previous ventures in the realm of dare-devilry.

Many times when he went to the Methodist Chapel and heard a preacher say something which he thought was not strictly in keeping with the old-fashioned Gospel of his choice, he had no hesitation in calling out with a loud voice: "Now then, lad, speak the truth and shame the devil."

The last time I saw Jimmy he was on crutches, and his greeting was, "I shanner want these things in Glory."

TUDOR ROSES GATHERED ON CANNOCK CHASE

It was a rare experience to gather Tudor roses on Cannock Chase. There was a pile of them lying in a corner of a seventeenth century building which was being demolished at Norton Canes. They had been detached from the ceiling of the main room of Norton Hall, which had become so seriously damaged by mining operations that it had been sold to be razed to the ground.

A stone that faced towards the Watling Street had on it the inscription : "Anno Dni 1682. Fr : Hanbury," and the ends of the iron spouting at the eaves had lions' heads at the corners. Already the work of dismantling had been proceeding for some days, and it was not an easy matter to explore the thirty or so rooms and to dive into the depths of the rambling cellars.

There was eighteen-inch brickwork everywhere, and the profusion of sound oak was astonishing. Great beams and towering rafters looked as clean and fresh as if they had been fixed in the last year or two. The staircase, with its wide oak steps and balustrade of patterned iron, displayed a curious mixture of design, but the oak panelling in all the rooms was uniform in its patterning and the dignity of centuries. The oak shutters to the windows and to the main entrance door recalled the day when dwellers on Cannock Chase had particular cause to shut themselves off completely from the outside world.

Massive slabs of thick stone paved the entrance hall, in which the remnants of a huge fireplace remained. There were deep spaces behind the oak partitions.

"No, we have not yet found anything particularly valuable," said the house-breaking contractor. He produced a King George III. penny of 1780 and two other pennies of 1806 which had been unearthed. Behind a chimney-piece there was discovered a seven-and-sixpenny ticket for Walsall Races in 1830, and a lad on the job had as souvenirs several quill-pens that came to light from behind a mantelshelf.

Everybody working on the pulling-down was keeping open eyes for possible riches, and one or two of the men had even been exploring the wine bins in the underworld in the hope of discovering something good there.

"Look out, there!" was the cry of a man on the top of things as a big mass of brickwork came toppling over, and the cloud of dust that followed was sufficient to blot from view the reconstructed Fleur-de-lys tavern on the Watling Street, some distance away.

One of the dismantlers was specially interested in the "L" shaped iron hooks which had been knocked into the rafters at regular distances for the purpose of holding the laths, and he was praising a slotting method which enabled him to salvage every lath in a perfect state.

Wide gashes in the walls testified to the wrecking effects of coal-pits in the locality, and the steel tie-rods about the building indicated the determined efforts which had been made many years before to save the mansion from disaster. At last, the old home of the coalmasters had become the victim of the coal-getters.

For several centuries the Hanburys were associated with the Cannock Chase coalfield, and for a long period they were owners of Norton Old Hall. In the early years of the nineteenth century William Hanbury was resident there, and made great developments with the mines in the Brownhills and Norton areas. His son, also named William Hanbury, carried on the succession, and a few years later there went to live at the hall the William Harrison who was one of the most remarkable men the Cannock Chase coal industry has ever known.

He was followed by William Harrison, who, in turn, was succeeded by William E. Harrison, and he and William Harrison the fourth—father and son—have maintained the family traditions in worthy style.

One of the rooms at Norton Hall still bore

evidences of having at one time been used as a nursery, for on the walls was a charmingly decorative paper which told dozens of nursery tales in rhymes which were brief and quaint. The oldest inhabitant said the rhymes were on the walls at the hall when he was a lad.

Not a sign of electricity was to be found in the whole building. Ashes told their story of Cannock Chase coal loyally used, and as the walls were falling about my ears I read the nursery news of real coal days when:

> "Little Polly Flinders
> Sat among the cinders
> Warming her pretty little toes."

MARY, HILDA AND MADELINE

In the Hungry 'Forties when Cannock Chase had no railway nearer than the Grand Trunk line which touched its western borders through the Vale of Penk, Mary Audley, the heroine of Stanley Weyman's "The Great House,'' alighted at Penkridge station to make her way over the moors of Cannock Chase.

Peter Bassett was Mary's companion, and they were the only passengers to leave the train at Penkridge. The platform was scantily illuminated by a single oil lamp which looked down on a dim churchyard. Dusk was passing into night, and the wind, sweeping across the flat, whipped Mary's skirts and chilled her blood. A light or two in cottage windows told of the nearness of the village, but in every other direction dull lines of willows or pale stretches of water ran into the night.

When clear of the village Mary, peering out of the fly which had been sent by her uncle to meet her, fancied they were winding through woods. They jolted into ruts and jolted out of them again. By and by they were clear of the trees, and the road improved. The moon, newly risen, showed Mary a dreary upland, bare and endless, here dotted with the dark stumps of trees, there of a deeper black as if fire had swept over it and scarred it.

They met no one, saw no sign of habitation. To Mary, accustomed all her life to streets and towns, the place seemed infinitely desolate—a place of solitude, witches, terror and midnight murder.

"What is this?" she asked shiveringly, and Bassett replied, "This is the Great Chase; Riddesley on the farther side, is our nearest town, but as soon as the railway was opened we used Penkridge station."

The gaunt sign-post that looked like a gibbet and might have been one, the skeleton trees that raised bare arms to heaven, the scream of a dying rabbit, all conspired to make Mary feel a passing depression.

Another mile was covered and Mary reached Beaudeleys Park, where she was to spend many happy days. She had glorious walks in the park, where she was able to follow the silver streak of the river winding afar between park and coppice and under many a church tower. Away to the right she could see the three graceful spires of Lichfield, and southward, where an opal haze closed the prospect, she could imagine the fringe of the Black Country, made beautiful by distance.

Mary came to know the Great Chase intimately, and almost daily she walked to a point beyond the park where she could see

BIRCHES VALLEY.

the distant smoke of a mining village. That was the one sign of life on the Chase; elsewhere it stretched vast and unpeopled, sombre under a livid sky, smiling in sunshine, here purple with ling, there scarred by fire—always wide under a wide heaven, raised high above the common world.

In the pages of "Casting of Nets," Richard Bagot describes the highlands of Cannock Chase as seen from the spacious meadowlands which extend between Abbotsbury and Trentford—Abbots Bromley and Rugeley. Hilda Redman it was who was introduced to new friends at Abbotsbury, where the old-fashioned garden had its sweet-scented flowers and herbaceous borders and a glorious view over the Trent Valley to the loftier lands of Cannock Chase. Hilda became the mistress of Abbotsbury's stately home, and she soon heard glowing stories of the sumptuous hospitality dispensed by her predecessors on the occasion of Twelfth Night festivities, culminating with a ball for the tenantry and principal workmen of the estate.

The barons' hall blazed with the light of innumerable wax candles; garlands of holly and mistletoe adorned the dark oaken roof and twined between the massive beams and rafters, while King John, pen in hand, with the scroll of Magna Charta before him, scowled down from the painted ceiling upon the varied throng of dancers on the floors beneath.

In the great gallery, the massive painted windows were uncurtained, and the light from inside the gallery shone through the stained glass, casting patches of colour on the snow beneath them. Three large fireplaces, in which were blocks of Cannock Chase coal and chunks of wood, warmed it, and the smell of the burning wood mingled pleasantly with the scent of the orchids, lilies of the valley—which grew wild in the park—and other flowers distributed about the room.

Twentieth century Hednesford makes its appearance as Hendiford in "The Wine of Illusion," written by Bruce Beddow, a native of the town, who introduces to the wild free Chase a beautiful gipsy girl, Madeline Lucian, who is full of strange allurements. Madeline climbs a Chase hill crest, whence she sees the far horizon of the north, dark, frosted blue in colour, and in another direction she sees upon a cleaner sky white clouds blowing out like galley sails in the breeze. Heather and gorse awaited the summer in faded quiescence, but the turf was full and springy after the fires of autumn and the winter snows. Five lean race-horses from Coulthwaite's stables trod the horizon in a sedate string. There was a whistle of the wind over all, and like a bass undertone to the whistling of the wind was the distant rumbling of pit-engines, rising and falling in seeming harmony.

THE LAST OF THE FLEUR DE LYS

The cry of "Time, gentlemen, please," of the landlord of the Fleur-de-Lys had a special ring about it, and the significance was clear on the following morning when the words: "Closed, Closed." were to be seen written in big chalked letters on the main doors of the inn.

The seventeenth century coaching house on the Watling Street had for many years been fortified with a thick iron waistcoat, and this had been supplemented by stout oak supports shored to the external walls. Gaping gashes had made their appearance in the structure, and the floors and ceilings had assumed sloping oddities. In the records of Cannock Chase the Fleur-de-Lys figures prominently. It was the favoured rendezvous of the parish fathers of Norton Canes and Little Wyrley in the eighteenth and nineteenth centuries. Its commodious rooms, with low oak-beamed ceilings, were the meeting-place of generations of natives, and often when worship was over at the neighbouring parish church the gentry of the neighbourhood would adjourn to "The Luce" to have a refresher, and, remotely possible, to discuss what the Rector had been talking about.

The inn was a popular calling-place for travellers on the great highway from London to Chester, and in the days when it was customary for cattle to be driven on foot from Holyhead to Birmingham and beyond, the drovers put up for the night at the Fleur-de-Lys.

Later, when the coal measures on that side of the Chase were mined by the Marquis of Anglesey and William Harrison—"Buttermilk Bill" he was nicknamed, because of his good-heartedness to his men—there was no lack of brawls and rowdyism in the crowded inn when migrant miners from Darlaston, Bilston, Portobello and Moseley Hole gathered to drink riotously.

One day when I went to the inn I found the greater part of the premises had fallen a victim to the house-breakers. Three pennies, dated 1779, and a couple of halfpennies of 1761, was the extent of the treasure discovered in the wreckage. Under one of the floors had been uncovered hundreds of pins of the old-fashioned kind—big, needle-pointed, on which the heads had been separately fixed. As one of the men shattered the brickwork of a chimney-breast he said to me, "I wish I could be as lucky as the fellow that found those thousands of pounds in the secret hole at Pillaton Hall."

From behind a bench in the bar I recovered an ace of diamonds, and it set me thinking of the customer who had slipped it away momentarily and had lost it.

The old sign of the Fleur-de-Lys had gone into safe keeping at Little Wyrley Hall, with whose family of Fowkes, Husseys and Macphersons the inn had been associated for centuries.

Whatever may be the correct pronunciation of Fleur-de-Lys there is no doubt those who live in the villages around the Watling Street inn call it the "Flower de Luce," or more frequently, "The Luce." Anyone who ventures on the "Fleur-de-Lee" version is viewed as a pedant, and I came across one Shakespearean student in the locality who claimed the famous dramatist as a supporter of "Flower-de-Luce."

When the new inn was opened in celebration manner, I found the miners of the locality talking animatedly about the event, but, disregarding all pronunciation they were exchanging views about the "Luce" opening its doors once more.

OLD SPORTSMEN WITH LONG MEMORIES

The finest pair of Cannock Chase sporting tale-tellers I have ever known were William Drury and Noah Corbett, and it was a day to be remembered when I succeeded in getting them together and setting them talking in reminiscent mood.

Natives of Shropshire—one of Hadley and the other of Lawley Bank— they had spent all their days since boyhood on Cannock Chase. Corbett was landlord of the Cross Keys, Hednesford, and the room at the hostelry in which we had our chat had been the haunt of countless notabilities of sport.

Drury was the older man of the two, and he was able to go back to the 1840's when Cannock races were a great attraction for many miles around. One year a horse named Intrepid ran at the local races and gained a victory which was the beginning of a long series of successes on the turf. Another year Drury had the mount of Pactolus, which won a good prize for George Perry, of Brocton Gate.

Drury was one of a large company of spectators who saw Frederick Pratt, of Saredon Mill, ride, for a wager, a horse named Spectre from the Swan Hotel, Wolverhampton, to the Greyhound, at

Yarlett Hill, Stafford, a distance of twenty miles, well under the hour. For the entire length the road was covered with ice, and at times the going was exceedingly difficult.

Drury knew Fred Archer very well, the famous jockey frequently visiting Hednesford to see his father, William Archer, who was at Bradley's stables. Drury and Archer were out together on the hills when the latter was having trials with the horses.

Only once in forty years did Drury miss seeing the Chester Cup race run, this being a record of which he was very proud.

There was no incident which gave him more pleasure to describe than that of his first meeting with Dr. William Palmer, the notorious Rugeley poisoner, when the drinking of the unlimited quantities of champagne helped to fix the day in his memory.

It was on the second day of Lichfield races. Drury and his friend, George Lovatt, of the Four Crosses, Hatherton, went early in the morning to see what horses had won at Lichfield on the previous day. When they arrived at the inn they found Palmer and William Saunders, his trainer, there. Palmer's horse, Polestar, had won the Staffordshire Stakes at Lichfield on the first day, and being in jubilant mood, the Doctor treated all the company to champagne. Palmer and Saunders then set off for Lichfield again.

Drury smilingly gave a smack of his lips as he remarked to me, "I've seen as many as a dozen bottles of champagne on the table at a time at the Keys in those days."

This reference to William Saunders led Corbett to chime in with a recollection or two of the noted Hazel Slade trainer. He recalled that Saunders won the Cambridgeshire in 1872 with Playfair, the jockey being George Ashworth, who also carried off the Shrewsbury Cup with Indian Ocean and the Liverpool Cup with Wynyard. His head lad was named Metcalf, whose nickname was "Badger."

Thomas Osborne, who rode Marie Stuart to victory in the St. Leger, was a native of Hazel Slade, and was apprenticed to William Saunders. Corbett also knew Robert Denman, whose training establishment was at Littleworth House, near Hednesford, and it was there that Robert Denham, junior, trainer of the French horse, Sir Galahad the Third, the winner of the 1924 Lincoln, was born.

"I believe I am the only man now living on Cannock Chase who saw the Chester Cup won by Goldfinder, which was trained by William Saunders, for Dr. Palmer," said Drury, and as he spoke the words he lingered on them with evident pride.

"I shall never forget," he continued, "the jollifications we had in this very place when Palmer and Saunders and a crowded company of friends met to celebrate the Chester victory. Speeches were made by the Doctor and his trainer, and the best part of the drinking was when Palmer treated everybody to a good supply of brandy. Why, there was as much brandy drunk that night as would sink a boat."

The talk veered round to cockfighting and coursing, and Drury became quite excited as he described mains which were fought at the rear of the Crown Inn, at Cannock, of which his father was the landlord. As a young man Drury had been in part ownership with Saunders of a greyhound named "Wee Nell," which won the champion puppy stakes at Newmarket, and was leading prize-winner at Southport and Ivetsey Bank. Drury won the Watford Gold Cup with Jacob, this being an achievement which he claimed to be an especially fine one.

This recital by Drury was followed by Corbett's enumeration of some of the scores of rabbit-coursing successes of his "Barbette," a brown whippet, which was for some years champion of the world. On one occasion the dog had a match at Wakefield for £50 a-side, and he won by eleven to none, getting the rabbit every time. Cherry Nose and Little Eremon were two other world challengers bred by Corbett. The former never lost a match, and the latter's name was chosen because Corbett's son-in-law, Alfred Newey, won the Grand National for Coulthwaite on

Eremon. Little Nigger and Bet were other frequent winners of whose records Corbett was justly proud.

Both Drury and Corbett knew George Kent who was a stable-lad at Flintoff's stables more than a hundred years ago. It was Kent who used to walk the horses to the race meetings all over the country. He was only ten years of age when he left his home at Aldridge and went to Hednesford, and in the same year he walked a horse named Birmingham to win the St. Ledger.

"George Kent was here at the Keys with me on the night that Palmer treated us all round with brandy when he won the Chester Cup. Almost the last time I saw Kent, when he was ninety years old, he asked me if I remembered the Doctor saying when the brandy was being passed round, 'If ever I live to win the Chester Cup again, I will treat you all round with brandy again, but God only knows.' " So said Drury.

"Which reminds me," added Corbett, "that my father walked from Dawley Bank to Stafford to see Palmer hanged."

"But how about my job of showing thousands of curious visitors the room in which Cook, Palmer's victim, died," commented Drury. "I kept the Shrewsbury Arms, at Rugeley, for eight years, and although it helped the business I can tell you I was dog-tired some days showing folks the room and telling them all about it."

CANNOCK'S HARVEST OF SPARROWS' HEADS

There was a great mobilisation in the quiet township of Cannock before the thrills of Waterloo had faded away, and everybody, from the perpetual curate to the smallest nipper in the parish poor-house, was set to work to bring sparrows' heads to the village constable.

"Do not bother your heads about it being nice up Shore Hill, or on the Chase," said the parish fathers in vestry assembled, "but look on the crops that are being ruined by these wicked pests, the sparrows, and set about slaughtering as many of them as you can."

This was a call to warfare which had a monetary inducement at the back of it. The overseers and the men about them discussed the rate of pay at no small length in the narrow confines of the chamber at the foot of the tower at the west end of the church, where the walls are crudely panelled with boards recording the generosity of former parishioners.

The uneven top of the gaunt, iron-clamped ancient oak chest the clerk of the vestry used as an emergency table on which to rest his minute book. Resolutions and amendments had to be subjected to friendly scrutiny before the decision could be reached that fourpence was to be paid for a dozen sparrows' heads taken to the constable. It was agreed that a proportionate allowance should be made for divisions of a dozen.

There was no need for the news of the mass attack to be announced either in the church, on the south door, or in the Market Place by the bell-ringer. The speed with which heads of victims were laid on the steps of the preaching cross in the churchyard for the constable's inspection was ample proof that the parish had responded readily to the "Save the crops" call. Old and young alike held out open hands for sparrow pence.

John Weetman was one of the greatest enemies the sparrows had. He kept a hairdresser's and cutler's shop, and so often does his name appear in the day by day records of the constable that we are forced to conclude that heads, human and feathered, must have absorbed the greater part of his attention. The passion must also have extended to his family, for we read of Weetman's boys having sums ranging from twopence to a shilling, and to relieve the repetitive monotony of his entries the constable tells of his having paid eightpence to the barber's boys.

The village blacksmith was such a familiar character, that William Thompson, junior, the constable, did not trouble to mention his name, contenting himself by stating his occupation, and the fact that he pocketed eightpence for fifteen heads. The captives had most likely been bagged when they ventured inside the smithy door to pick up the corn which had fallen from the nosebags of tethered horses.

In those days the Rev. William Blow Collis, M.A., did duty in several Chase parishes, and that even he was not averse to augmenting his income by means of sparrow money is seen in the acceptance by him of fourpences for a number of dozens of heads. "Richmon Collis, eight sparrows' heads, threepence," reveals that the curate's boy had a little pocket money for useful service rendered to the village.

When the name of "Mr. Masfen" comes into the constable's book as the recipient of head money we realise that the landed gentry of the locality were likewise ready to come on to the rates. The constable had a special spell of book-keeping to do one June day, when the birds of Cannock must have gone a-sighing and a-sobbing. One Trupshaw deposited thirty-one heads at the feet of the constable and received tenpence; Juke's victims numbered a dozen, and others who that day shared in the disbursement of sparrow tribute were Weetman, Turnock, Lewis, Amos, Shaw, Sambrook, and several lads, one being distinguished as "the lad from the mill."

One of the unnamed sparrowmen was "the horseler at the Crown;" other entries tell of "Brindley's girl," "three boys from Rumer Hill." Pickerill and Cooper had good scope on their lands for bird-shooting, and we find that bags of six and five dozen were often credited to them in the constable's book.

In 1819 malcontents in the parish started an agitation in favour of the payment of sixpence instead of fourpence a dozen for the heads. Pending the settlement of the issue by the vestry, the constable took upon himself the risk of conceding the extra amount in respect of a total of six dozen heads. When the time arrived for the half-yearly accounts to be approved by the overseers one shilling was deducted from the constable's wages "on account of the sixpence per dozen charged for sparrows' heads, the parish allow fourpence per dozen."

Collis the curate was the first in the parish to get the threepence which the vestry resolved to pay for the bringing of an "edgog" to the constable. Three edgogs, ninepence, one dozen heads, fourpence, total one and a penny," is a typical entry. Such payments continued for over twenty years, and in the parish books the items appear among a medley of other entries relating to payments made to men and women on the roads, old soldiers making their way from place to place with passes, and the parish poor in the workhouse.

IN FOOTSTEPS OF ANCIENTS IN BEAUDESERT PARK

Having circled Castle Ring one hot August morning, I dipped down into the great Beaudesert Park, bent on rambling at will where footpaths did not exist. "It is the finest piece of ground of its kind in all England," said a noted Duke of Bedford about Beaudesert, and before the day I had not the least hesitation in accepting the declaration.

It was a delightful experience to drop lower and lower until I reached the stream several hundred feet below, into which a score of tributaries tilted their spring waters. These musical rivulets emerged from hollows where rabbits came to have their sips, and they made their way beneath dense bracken barrages.

The rambler must have an ear for ripples if he is to escape being tripped up in the intersecting courses. Added to the sounds of the hill-side rills were the songs of innumerable birds which flitted from tips of young trees to spikes of sturdy undergrowth, the clack of grouse as they struck out towards a distant copse, and the incessant buzz of myriads of flies whose attention I had to combat by a continual swishing of a bracken frond.

There had been much oak-felling and many gaunt stumps marked the sites of vanished veterans. Lying amid the bracken was a decayed, overthrown oak whose recumbent ribs towered in front of me to a height of a dozen feet. Efforts had been made to fire the wreck, a black, ten-feet long cavity showing where the flames had bored their way. On top of the ruins were several bracken seedlings and red willow-herb plants, the product, doubtless, of seeds flung by the wind.

Here and there, I came to the rim of an Elizabethan bell-pit, and close to a group of these I found pieces of ironstone-slag scattered about. My immediate thought was of the ancient smelting furnaces which formerly were busy in many a hitherto secluded corner of the Chase. Following a narrow track for some distance I came to the conclusion that it was most likely the way by which coal was conveyed from some of the bell-pits to the Paget mansion, where well into the nineteenth century all the household fires were fed with fuel from the park seams.

A considerable amount of material had been laid down for the strengthening of the track, and from many conglomerates of broken earthenware and tiles I salvaged several particularly intriguing fragments.

Ahead of me was the silver surface of the Trout Pool, whose waters had proved to be such a precious resort when the fire-fighters had to battle for hours with the flames which threatened to destroy Beaudesert Hall on Guy Fawkes Night, 1909. By the side of the Trout Pool I found a flourishing brother of the noted Rawnpike Oak, whose fate was sealed when it was struck by lightning in 1932. The pool-side giant was a magnificent specimen of woodland beauty, and it was commendably good sense on the part of the fellers to spare it when the park was cleared of nearly all its oak in the 1920's. Declared by expert woodmen to be not less than eight hundred years old, the oak was about eighty feet in height, and its branches had a measurement from tip to tip of one hundred and twenty feet. The tree was sound in every part, and marks on its sides showed where a woodman making a test had found the timber to be as tough as his steel.

The Horse Pasture, hereabouts, had once a great crowd of fine oaks, for the subsoil is rich with white loam of many scores of feet thickness, and the timber produced here has always been of the best. On the slope where the Horse Pasture runs up to have contact with the outer earthworks of Castle Ring was a small brick structure, used many years ago for the feeding of the deer. So frequently were the deer to be seen at this spot that Deer Slope was the name given to it. Fifty years have gone by since the Paget keepers last fed the animals from the small building on the

Deer Slope, but almost on any evening the deer may still be seen going to the discarded feeding-place. Is there any strong instinct or tradition persisting in the herd?

At the end of a long valley there was a disused rifle-range where the Pagets, always noted in military annals, kept themselves for generations in trim ready for service on every British battle front.

As I made my way back towards the heights of Cannock Wood, I passed a quarry from which was obtained the stone for the erection of Beaudesert Hall centuries before, and also for the extensions to the mansions in 1914. A number of churches in the Cannock Chase district had likewise been constructed of material from the park quarry.

My last park impression that day was of a remarkable tree which stood by the track-side not far from the quarry. It was a beech, lofty, perfectly straight, and had not a solitary blemish of vein, scar or effusion on the whole trunk. It was a handsome Venus of the woodland.

THE VILLAGE BLACKSMITH WAS FIRED FOR SCIATICA

Thomas Fletcher, the village blacksmith of Walton-on-the-Hill, claimed to know every inch of Cannock Chase from Baswich to Pye Green. At the age of ninety-three he was able to dig and plant his seven rood garden, and with corresponding freshness of mind could recall clearly lots of things that happened more than eighty years before.

In his boyhood there were thousands of sheep grazing on the Chase. Quite a number of men had eighty to a hundred sheep each on the moors. The animals made their own walks and kept inside their self-imposed boundaries with remarkable faithfulness, so that their owners knew exactly where to find them.

Cannock Chase sheep always fetched top prices at the markets, and men came from all parts of the country to buy them. Deer abounded on the Chase, there were plenty of blackcock, and he could remember the time when he caught trout in Sherbrooke stream.

It was a delightful experience to be out of bed at three o'clock in a morning and go bilberry picking on the moors as soon as it was daylight. Sometimes it was just possible to see the berries on the wires in the half-darkness. The pennies gained by selling four or five quarts of bilberries were riches to him in those days.

Thomas was born at Penkridge in 1843, his father being a coachman for the first Lord Hatherton. He aspired to become a blacksmith and wheelwright, and Lady Hatherton made herself personally responsible for the premium paid when he was apprenticed at Acton. After eight years at Acton he moved to Walton, and there he had continued to live for sixty years in a cosy cottage near the village pound and the quaint thatched post-office.

He admitted he had been lucky in his general health. Once when he was a young man he was fired by the village doctor for sciatica, but that was only just four dabs of a poker which had been well prepared in the smithy fire, and that was the last he ever knew of the complaint.

The next worst thing he suffered was when he kept bees and he was nearly killed by a sting during a swarm. He lay at death's door for days, and then he did with the bees as he had done with the sciatica : finished with them once and for all.

Of course, there had been various minor accidents in which he had been concerned.

"Let me see if I can remember a few of them," said he, and he proceeded to count the sixteen times he had either been thrown out of vehicles or knocked down in one way or another.

"But you see, I never had any bones broken, and that's what counts after all, isn't it?" he added.

Established in the ingle-nook of his two hundred year old cottage, the blacksmith told me of his farriery successes at the County Agricultural Show, one of his wins being a first in a field of twenty-six at Burslem. A gold-plated horse-shoe hanging on the wall was a treasured proof of the skill which could make a new article out of the iron of a cast-off.

One day during the great military manœuvres on Cannock Chase in the summer of 1874, Thomas was watching the Seventh Lancers galloping on the hills near Brocton Gate, when a shoe of one of the horses went hurtling through the air. The blacksmith quickly retrieved the cast-off, and in five minutes it was soundly re-fixed, to the surprise and delight of the Lancer.

In keeping with his love of the ring of the anvil Thomas had a fondness for the clang of the bells, and so for fifty-five years without a break he was a ringer at Holy Trinity, Baswich, the mother church of the parish. A tablet in the church commemorates his completion of fifty years of service.

He took part in more than one merry Jubilee peal, but the villagers thought Thomas never did his work with more gusto than on the Sunday after his notable Burslem win, for he was immensely proud to have "put Walton on the map."

THE RIVER GOD OF CANNOCK CHASE

One day I went in search of the "River God of Cannock Chase." I failed to find him, for time had carried him away. He used to be a very fine leaden figure and had his place of honour on a bridge in Hagley Park, through which the waters of the Chase run before uniting with the Trent.

"Old Cannock Father of the Seven Springs" was the inscription beneath the figure, and it was supposed to allude to the number of the streams that flow from the Chase into Hagley.

Old Bob Curzon took a special pride in the River God, and it was said of him that he soon put up his spying-glass to his eye if he caught sight of any intruder in the park who was likely to scratch his initials on Old Cannock's form. So fond was Bob of his spying-glass that he was nicknamed Paul Pry.

Mr. James Moore, who lives at the edge of the park, told me he had seen the River God on the bridge many years, but he did not know what had become of it. It had a narrow escape one day in 1874 when Cannock Chase was visited by what was described as a water-spout. All the dams of the pools on the moors and in the Rising Brook valley were broken down, and a huge volume of water rushed through Curzon's Park. It swept to the ground a long stretch of the ten feet high park wall, and, continuing its mad career into Rugeley, tore up the streets, made deep chasms in yards and gardens, demolished lamp posts, and washed away the entire stone pavements in a number of the streets. A great number of houses were flooded, and furniture was born away on the torrent.

That was an occasion when, for once in a way, the picturesque phraseology of a contemporary topographer became an anachronism: "the park is obscured from the vulgar gaze by the park wall and fencing."

Hagley Hall had its beginning in the time of Edward III., at which period the estate was in the hands of the Westons, whose name is identified with Weston-under-Lizard. Sir Simon Weston was Sheriff of Staffordshire and Recorder of Lichfield in the reign of James I. Another member of the family was Sir Richard Weston, who was created Earl of Portland by Charles I.

There was a Robert de Curzon who was reputed to have come over with the Conqueror, and it was in the eighteenth century that Hagley Hall came into the possession of Assheton Curzon, second son of Sir Nathaniel Curzon, fourth Baronet, of Kedleston, Derbyshire. Assheton Curzon was created Baron in 1794, Viscount in 1802, and his death took place in 1820, when the peerage expired.

Hagley was used as a farmhouse when the Curzons took it for use primarily as a hunting-box for Cannock Chase. Extensive additions and improvements were made to the hall, and when the Hon. Robert Curzon married Lady De la Zouch and lived at the hall, Hagley became the centre of many brilliant social gatherings. Hospitality was dispensed with a lavish hand, and with Beaudesert, Lysways and Bellamour also flourishing as great houses, Rugeley had a good share in the district's distribution of wealth. All these mansions, alas! have been felled, either entirely or partly in recent years.

Bob Curzon formed a Literary Institute in Rugeley in 1850, and after the fashion of the time he renamed the society some years later as "The British Workman." This was a revision which brought more of the Brereton colliery workers into membership, the "workman" touch winning where "literary" deterred. Curzon was one of the governors of Brewood Grammar School, and was badly bored one day when forty-two applicants for a post of assistant schoolmaster had to be interviewed. He arrived back at Hagley in a dead-tired condition.

When the Curzons ceased their ownership of the estate the hall passed into the hands of several men one after another, and one of them, Mr. E. W. Colt, claimed to have recorded the appearance in the park of eighty species of British birds.

Only a fragment of the mansion now remains. The River Godless bridge spans the stream; the pools retain much of their beauty, and the high brick wall surrounds the park on the town side, but there are many acres of park land without a stately tree to dignify it.

I have seen Hagley Park in many moods on days far apart, and the picture which remains most vividly in my memory is that of the summer day in 1914, when, at the urgent call of duty, the thousands of young men of the Officers' Training Corps hurried away from their camp to take up sterner tasks in Britain's cause, and a great number of them to die on the battlefield.

SID'S ENGLISH OF THE SONOROUS ROLL

Many people have wondered what became of "Sid" Sharpe's barrow. If, when Sid ended his days, there had been a Cannock Chase museum for preserving things linked with pleasant memories, that barrow would certainly have been found a place therein. "Sid" was an out-door porter without official attachment, and commercial travellers had a special affection for his spicy personality.

Witty, droll and pungent in his expressions, "Sid" saturated his services with volubility. When waiting for something to turn up, he would sit somnolent on his barrow, his arms folded and his head nodding at intervals approval of his hidden views of men and affairs.

If he had received a post-card addressed: "Sidney Sharpe, Esq., The Station, Hednesford," requesting him to meet a certain train, you might see him on one of the platforms in a livelier mood. Moral reflections would trip from his lips by the score; poem after poem, grave and gay, he would roll off with ease and perfection of speech, and for every person who came along he would have a salutation as to health, weather, trade outlook, holiday hopes, local events, or domestic interests.

Sometimes he would rest with his cargo in the busy main street. Instantly, a conversation would spring up between himself and a shopkeeper or a passer-by. His garrulous chatter would swell into louder tones, and before many minutes had gone by a group of people would be gathered round listening to "Sid's" vivacious monologue.

"Sid" had a superior touch about him. He made a speciality of the English language; the English language of the sonorous roll. "I will make investigations about the best facilities for your transit," he would say to anyone who was in doubt about a train-time.

"Profusion of vehicular traffic has impeded my progress," was his way of explaining why he was a little late in calling for a trunk or a hamper.

"My word, what a conglomeration of holiday necessities," he would say aloud to himself as he loaded his barrow at the house door of the intending traveller.

"Chronological promptitude has always been my motto"; "I always endeavour to coincide with your kindly injunctions"; "My aim is to merit your renewed patronage, and win your approbation"; "No one can dispute that 'Sid' tries to be indefatigable and conscientious in strict adherence to duty," were some of the fond phrases that were often used by him.

Sid, I want some luggage fetching for the ten train in the morning," was my remark to him one evening.

"At your obedient disposal, sir. I shall be honoured to execute your orders with expedition," was his reply.

The luggage was duly called for, and as he wheeled his barrow down a slope I offered a retarding hand for the load.

"It is quite equipoised. I can negotiate the declivity with facility," was his assuring intimation.

On my arrival at the station, "Sid" informed me he had affixed the label. While waiting for the train, he gossiped about the kind of weather the wind was promising, and prophesied that "the sun will continue to illumine the earth for at least a week."

"Wales! Wales!" he murmured; "that is a land I have never been affluent enough to traverse, but Welsh mutton is sumptuous to the palate. When I eat it its succulence makes me imagine I am a prince of fabulous wealth, so luscious is it after my ordinary, meagre meals of plebeian fare."

His last words as the train steamed out were: "Now if you desire me to convey your luggage to your abode when you return, please despatch me a post-card to that effect."

When "Sid" Sharpe went on his last long journey, and his voice was no longer heard in the streets, Hednesford mourned for many days the loss of his breezy Johnsonese.

POETRY AND PROSE IN BLITHEFIELD COURTYARD

There was a time when prime treacle beer could be had at a cottage which stood on the Chase not far from Pottal Pool, and when the Meynell, South Staffordshire and the Albrighton Hunts met at the spot there was always a rush by thirsty ones who wished to sample the special tasty beverage.

Another noted Cannock Chase drink was on tap at the park keeper's lodge kept by one Turner, at Blithefield. It was described as an extraordinary ale which could not be equalled unless brewed with the water obtainable in that locality, and kept in the cellars of the lodge. When kept till old the ale mantled when drawn out of the barrel, and it was declared by ale tasters to surpass description.

Henry Luttrell, poet and author, drove over Cannock Chase one day in the 1840's to sample the Blithefield prime, the journey being made in the first brougham ever seen in Staffordshire. Having indulged in Turner's brew Luttrell declared himself to be in the third heaven, and, in accordance with his literary tendency, broke into rhyme :

"Walk where I will, or ride, or drive or sail,
I ne'er shall see such oaks nor taste such ale."

When, in the course of a circuit saunter from Rugeley to Colton, Blithefield and Abbots Bromley back to the Trent-side town, I called at the lodge where Turner used to assist visitors towards ecstatic moments, I saw several of the old mugs from one of which Luttrell most likely had his nectar. Later in the afternoon I saw in the courtyard at Blithefield Hall a number of relic vehicles which shared in the jaunt with the county's pioneer brougham.

The massive family coach, of unrelieved black, was there also. It had been built in sturdy style to withstand the rough passages through the park for shooting parties and for journeys to race meetings up and down the county. With the coach were a landau and a phaeton, reminders of the gentler pleasures of slow locomotion days.

Piled up about the wheels of these survivals from the past century were numerous minor personal belongings of the fourth Baron Bagot, which were to be offered for sale. Bagot oaks and acres had been lately sold for nearly £100,000, and here were several hundred unpretentious oddments which might be the means of adding a few more pounds to the total.

No bids were invited for coach, landau or phaeton and other out-of-fashion wheelers. They had simply been drawn out of the coach houses to make temporary room for the staging of a few score miscellaneous lots. At the end of the day they would be pushed back again into their idle but not ignored obscurity.

Moving about the courtyard among those who had come from many miles around was the fifth Baron Bagot, and it was he who set the sale in motion by bagging a bundle of umbrellas for five shillings. They were a motley collection, a colourful feature being a huge tent-like bobby-dazzler which had done duty on party days on the Blithefield lawns. The Bagot love of sport was well in evidence, but with the exception of a special rifle which scored a bull's eye by fetching thirty shillings, none of the many guns sold for more than five shillings. A pair of skiing boots went for the proverbial price of an old song, and bundles of fishing rods, riding whips, dog whips, walking and mountaineering sticks were almost given away for the asking. Two pairs of antlers of Bagot's Park deer were knocked down for eighteenpence; a mountain of commodious travelling trunks was conspicuous for the broad bands of yellow painted round them, and a loud hit-you-in-the-eye "B," also in yellow.

A shield bearing the Bagot arms was the central design on a number of chairs which caught the fancy of the high-elect bargain seekers, and ormulu clocks and Imari vases attracted the highest bids of the day.

There was a quick climb-down when a paltry shilling was all that could be secured for a terra-cotta bust of the Duchess of Sutherland, who was worshipped as one of the rarest beauties of her day.

Art in the courtyard that day was far different from the paintings of the great masters associated with the mansion for centuries. Haphazard landscapes and Egyptian scenes could not tempt anyone to express his shillings in double figures, and a pleasing forest study was needed to top the sovereign. Even a hunter standing with the Beggar's Oak as a background failed to reach the thirty shillings mark.

When it came to the turn of literature it was found that stacks of books which were reminiscent of ampler days of ease and time-killing immersion made small appeal to buyers. Forty volumes of Sir Walter Scott were in sedate brown leather, and their edges of lavish gilt dazzled in the sun as the auctioneer's man held a sample fistful for the crowd to see. Pope, Byron, Lytton and Herman Melville were complete in similar pomp of gold and leather, and it was Lord Bagot himself who had the favour of the buying bid for the eighteen dapper volumes of the author of "Moby Dick."

There was a heap of coloured, cardboard-covered novels, with an enticing variety of titles, which provided sugary sentiments and sensational spice to Mid-Victorian railway station bookstall patrons. Huge packages of the "Racing Calendar," Army Lists, two hundred volumes of the Journal of the Royal Agricultural Society—all these went in one lot for a solitary shilling. Fat Hansardian records of Parliamentary Debates, blue-paper covered copies of the Edinburgh Review, Royal Academy catalogues were included in the piles of lumber which no one seemed disposed to accept as a gift.

A queer mixture, indeed, was a corded bundle which held in compulsory companionship Samuel Butler's "Erewhon," a Court Guide of early Victorian times, four volumes of Tolstoi's hottest gospel, and a Book of the Peerage and Baronetage.

OLD TIMOTHY WAS SURE ABOUT GENTLE SHAW

Old Timothy Cumberledge, whom I saw one day looking over a gate by the wing-shorn windmill at Gentleshaw, had not the least doubt that the village came by its name in the way that his father had told him many years before.

Timothy's version was that a fellow named Shaw lived thereabouts a great while ago, and he used to go about the villages selling odds and ends of things with a pony and cart.

One day Shaw and his missis were going down the hill towards Longdon, the husband being at the animal's head, when the pony began to travel much too fast for the old lady. She shouted out: "Gentle Shaw! Gentle Shaw!" Thus, according to Timothy's oral tradition did the local place-name have its birth.

I found it was of little use to tell Old Timothy of a John Gentyl who lived in those parts six hundred years ago, and planted a grove or "shaw" of trees which became known as Gentyl's Shaw, "schawe" being a Middle-English word for a wood or plantation. Simple, on-the-surface, solutions of place-names have a firm hold on Cannock Chase, and so Old Timothy refused to throw over his father's homely story of Shaw and his pony.

It was Richard de Stile who, in like manner, set a spinney on another highland of the Chase, and provided us with Stile's Shaw, Stile's Coppice, or Stile Cop.

The Marquis's Drive reminds us of the "Waterloo" Marquis of Anglesey, who made the well-known track over the Chase from Beaudesert to Sherbrooke Valley. The Wellington Belt at Hatherton derives its name from a visit paid by the Iron Duke to the hall there, and the woods in which he had good sport.

St. Chad's Ditch tells of the limits of Lichfield Cathedral lands on the moors which in later times have been known as Chadsmoor. "The fosse of the Blessed St. Chad," was the ancient name.

Great Wyrley parish councillors were considering parochial boundaries one night, and as they moved their fingers over the map, which was produced from its long steel case, they came to Madman's Lane.

"Dunner thee know why it's called that? Why that's where Johnny Benton shot the devil!" said the father of the council to a younger member whose knowledge of parish history did not embrace past notorieties.

The grim, the eerie, the romantic and the charmingly rustic exercise their spell on many of the names of lanes, fields and woodlands on the Chase. Hangman's Oak is a wild, lonely spot to pass on a dark night, but

Goblins' Green need not upset our nerves. By Bear's Brook we need not be afraid, and in Hell Clough we shall come to no harm.

In Hart's Coppice and Deer Slade, by Deer Barn, on Hare Hill, in Cuckoo Bower and Cuckoo Lane we shall have pleasant experiences. In the Sweet Bit, Campion Wood, Mossy Lea, Cherry Orchard, Broomy Fields, Pear Tree Gorse, Dapple Heath Plantation, Birch Dell and Little Daisy Bank we shall find much loveliness.

Singularities which give distinction to the countryside can be found in Alder Leasowe, Rushes Meadow, Teasley Meadow, Flag Meadow, Broad Hazels, Stony Flats, Stoney Fields, Flaxland Meadow, Ozierbed Meadow, Rookery Fields. The personal touch greets us in Dan Brooke's Meadow, Jongham's Cottage, Wilkinson's Bower, Darling's Hayes, Biddulph's Pool, Sankey's Corner, Jacob's Hall Lane and Clergyman's End, the cryptic nature of the latter leaving ample room for surmise.

On the track of the beating of the bounds we find Gospel Tree and Gospel Well; and Coal Pitt Way, Colepit Field, Coalpit Leasow and Ironstone Road link us with the far-off days when coal and iron were first mined on Cannock Chase.

FAREWELL TO BEAUDESERT HALL

It was sunshine everywhere and not a cloud to be seen when I went on a July afternoon to bid "Farewell" to the mansion which had been the home of the Pagets for over four hundred years.

The lawns were not so trim as they used to be, and the gardens were devoid of their former glories, but the massive trees were as impressive and inspiring as ever. There was one touch of colour which seemed to be evidence of a decline from the cultivated to the unkempt : there were many spikes of red willow herb, wanderers from the wild, trespassing in more than one leafy bower.

Men were disturbing stone steps and crazy-pattern pavements; a large swan, with leaden wings outspread, was being hoisted from the lily-pond to be carted to another Arcady; beautifully embossed leaden spoutings and water-heads were being torn from gables; ladders by gaping spaces showed where windows had been bodily removed; ugly-looking lumps of iron which were being thrown through a window gap of an upper storey told of the smashing of fittings which was going on within.

The ancient guard room I found to have lost its former stateliness within a few hours.

WINDOW OF THE BANQUETING HALL, BEAUDESERT.

It was wreckage everywhere; stone floor and beautiful panelling all gone. The greater part of the twenty-thousand square feet of oak panelling in the other rooms of the mansion had likewise vanished with a speed akin to that of a migrant bird. Oak floors, decorative plaster ceilings, Jacobean over-mantels, fire-grates and irons of ancient dates, marble bathroom equipments were being dismantled by the bargain-makers' men.

In the Gothic banqueting hall the walls had been stripped of their massive panels. The Henry the Seventh carved oak top of the mantelpiece lay on the floor. Men were carefully removing the treasured minstrel Gallery, whose oak they declared to be in almost perfect condition.

Men with barrows were acting in accord with the terms of demolition that "waste material must be dumped in the hollow between the mansion and the stables." Loads came from hall and stables, and the quiet hollow of beauty was being turned into a haunted valley of tragic bygones.

For two days I tramped through the rooms of Beaudesert while the auctioneer disposed of the 570 lots, and secured thereby £8,000 for the whole of the fabric and fittings of the mansion.

Sixteen years before I had been present in the same hall when for ten days the valuable furniture was sold in 2,000 lots, and fifteen years still further back there was another Paget sale which raised great interest from one end of the kingdom to the other. On that occasion, thousands of articles of personal apparel of the Fifth Marquis of Anglesey were sold under the hammer. England smiled at the bejewelled garters, the rainbow tinted socks, the dazzling waistcoats, the 83 dressing gowns, and the 47 pairs of slippers —many of them golden and silken ones.

Three upheavals in little more than thirty years have been the fate of the Pagets of Beaudesert. Of the last, the Marquis of Anglesey has said that the feelings of his family and himself at the demolition of Beaudesert Hall could be readily imagined, and his regret was increased by the feeling of separation from Staffordshire friends. He had made every effort to find some use for the hall by means of which it could be spared, but nobody wanted it. Consequently there was no alternative to its demolition.

I mounted the "Waterloo" staircase—which had been left for the time for the convenience of mansion-breakers journeying with their loads—and entered the Long Gallery, whose panelled oak had remained from the days of Queen Elizabeth to those of King George the Fifth.

Standing in a wide window recess, I thought of Thomas, third Lord Paget, who in the days of Elizabeth stood on that very spot and surveyed the extensive lands around his "enlarged and new-re-edyfied" mansion.

A few yards from the Tudor doorway of the hall I could see the Castle Ring spring water pouring from the "Lion's Mouth" in as full a volume as ever it had done in the centuries of its flow.

Hurrying along the brown ribbon of road leading to the Gate House and the outer world were motor lorries concerned with the break-up of the great house of Beaudesert.

Beaudesert Park will still remain a place of loveliness, and all who enjoy the friendship of Cannock Chase sincerely hope that Lord Anglesey and his family will maintain the centuries-old tradition of the Pagets as friends of Cannock Chase.

AN ADDENDUM OF SUGGESTED ROUTES

The wide open spaces and the sheltered valleys of Cannock Chase can be approached from Cannock, Hednesford, Stafford, Penkridge and Rugeley. From the county town a frequent service of Red buses to Milford gives easy access to the moors and hills, and there is no lack of ways by which many beauty spots can be reached.

Sherbrooke Valley, Abrahams Valley and Seven Springs can be readily found without much consultation of maps. That is where Chase lovers in Stafford score over others who have to get to the bracken and heatherland by roundabout routes.

From Cannock a Red bus may be taken to Pottal Pool (three miles). There is a pleasant mile walk from there to Fisher's Lodge and the military cemetery. A return can be made by the same route, or *via* Pye Green, where a Blue bus travels to Cannock or Hednesford.

The military cemetery is a good landmark for several walks. One path goes to the Second Lodge, otherwise called Spring Slade, and the track to the left joins the Stafford to Cannock main road near Bednall Head.

From Spring Slade cottages a track to the right leads to the mouth of Sherbrooke Valley; the rough road straight on from

Spring Slade emerges at Brocton Gate on the Stafford to Cannock road; a path which strikes obliquely between the track to the Gate and the Sherbrooke entrance traverses the moors to the Messines Model, to Brocton and Milford. A right-hand deviation of the track drops down to the pools in Sherbrooke.

At the rear of the military cemetery there is a path which goes to the same mouth of Sherbrooke. From Fisher's Lodge a walk can be taken to the White House—now not quite as its name suggests—and to Range Corner.

What is now known as Range Corner was formerly the centre of Rugeley military camp in the Great War days, Smith's bookshop then standing on that spot. There are four ways at Range Corner. One goes through the gate and descends by the plantations to the mouth of Sherbrooke Valley. The track right ahead leads to Spring Slade. The second road from the Corner proceeds to the rifle ranges, beyond which the track continues down the ancient South Street on to the Rugeley to Stafford main road near Wolseley Bridge.

At the back of the sheds by the rifle butts a path to the left is the way through Old Brook Valley to Seven Springs and the same highway between Rugeley and Stafford.

The third route from Range Corner is the Penkridge Bank road to Rugeley. A mile down the hill a left turn goes to Etching Hill, and, as an alternative, the full descent of the hill may be made, and a turn taken by Stone House to the right to Slitting Mill, and Hednesford.

Another choice is to walk from the White House towards the foresters' houses at the head of Brindley Valley, then go right ahead, make a slow descent towards the L.M.S. Hednesford to Rugeley railway crossing at Moor's Gorse, where the main road between the towns mentioned is reached. A Green or Blue bus is available at that spot for either of the towns.

From Hednesford a Green bus provides a service down the Rugeley Road to Slitting Mill, where a walk can be had through the heather by the Fair Oak pools. The track climbs for a mile between the plantations to the top of the hill and contact is gained with the foresters' houses at the head of Brindley Valley, where a Green bus can be boarded for Hednesford.

Hednesford is the starting point for Blue buses for Cannock Wood and Beaudesert, and by this service Castle Ring may be easily reached. From Castle Ring there is a path through Beaudesert Park to the road which links Hazel Slade with Longdon and Brereton. At the cross-roads at Wandon Lodge one road passes over Stile Cop to Rugeley (two-and-half miles), and another dips through the valley to the Pumping Station at the junction with the Rugeley and Hednesford road.

Red buses from Stafford and Rugeley are useful to get to the Haywood Park end of Sherbrooke Valley, the lovely birch dell being within three-quarters of a mile walk of the Stafford to Rugeley main road. This is one of the most popular places for picnics.

A route by which it is possible by cycle or car to see a good portion of Cannock Chase has its start at Cannock. The direction from there is to Pye Green, the military cemetery, from there a short step-back to the cross-roads at Fisher's Lodge, then on to the White House and Range Corner. A stroll can be taken through the plantations inside the gate, and then a return made to the cross-roads. A move down Penkridge Bank brings Slitting Mill into view, then Sheepwash, where a left turn is made towards Rugeley alongside Coulthwaite's racecourse. A turn right to Stile Cop gains a fine view of the countryside. A return down the Cop to the road brings Rugeley at hand. Through Rugeley the direction is along the Stafford Road, past Wolseley Hall, along the bank of the Trent to the first bridge over the river. The track on the opposite side of the road to the river bridge is the entrance to Seven Springs. From Seven Springs Little Haywood is reached by the road over the river bridge; Great Haywood is reached a mile further on. Essex Bridge and Shugborough Hall can be viewed in the last-named village. A curving road out of the village goes past the ancient gate-house at Tixall and emerges by Milford station on to the village green.

The next move is to Brocton village, to the Stafford to Cannock main road, then on to Bednall Head and Pottal Pool, where a right turn is taken towards Penkridge. A left turn down a long narrow lane brings a junction with the Penkridge to Cannock main road; a passage over Shoal Hill, and Cannock Market Square is reached again a mile further on.

From Cannock and back to the same point by this route is about twenty-five miles.

In some instances bus services are not frequent—particularly between Rugeley and Hednesford, *via* Penkridge Bank, by Green buses—and careful consultation of tables must be made to secure the most profitable use of the facilities.

Some places mentioned in the Book

The figures denote approximately the number of miles

ABBOTS BROMLEY.
6 from Rugeley.

BEAUDESERT PARK.
1 from Moor's Gorse.
2 ,, Longdon.
3 ,, Rugeley.
2½ ,, Hednesford.
5 ,, Lichfield.

BLITHEFIELD.
4¼ from Rugeley.
3 ,, Abbots Bromley.

BROCTON.
4 from Pottal Pool.
1 ,, Milford.

CANK THORN.
3½ from Cannock.
¾ ,, Pye Green.
1 ,, Pottal Pool.

CANNOCK.
9 from Wolverhampton.
9 ,, Stafford.
8 ,, Walsall.

CASTLE RING.
3 from Hednesford.
1 ,, Beaudesert Hall.

COLWICH CHURCH.
3 from Rugeley.
3½ ,, Milford.

CONGREVE.
1½ from Penkridge.

DICK SLEE'S CAVE.
1 from Seven Springs (Colwich).

EDIAL HOUSE .
6 from Cannock.
2 ,, Lichfield.

ESSEX BRIDGE.
2 from Milford.

FAREWELL CHURCH.
2 from Castle Ring.

FOUR CROSSES.
2 from Cannock.
2¾ ,, Gailey.
2¾ ,, Four Ashes.

HATHERTON HALL.
1½ from Cannock

HEDNESFORD.
2 from Cannock.
5 ,, Rugeley.

HILTON PARK.
4 from Cannock.
5 ,, Wolverhampton.

LONGDON.
3½ from Rugeley.
4 ,, Lichfield.

MARQUIS'S DRIVE.
5 in length from beginning at Beaudesert Park to Sherbrooke Valley end.

MESSINES MODEL.
1 from Brocton.
3 ,, Military Cemetery.

MILFORD.
3 from Stafford.
7½ ,, Cannock.
5½ ,, Rugeley.

MILITARY CEMETERY.
3¼ from Cannock
¾ ,, Pye Green.
1 ,, Pottal Pool.

MOOR'S GORSE.
2½ from Hednesford.
2½ ,, Military Cemetery.
2½ ,, Rugeley.

NUN'S WELL.
3 from Hednesford.
½ ,, Castle Ring.

OAKEDGE PARK.
4 from Military Cemetery.
1 ,, Wolseley Bridge.
2½ ,, Milford.

PENKRIDGE.
4½ from Cannock.
3 ,, Pottal Pool.

PILLATON HALL.
3¾ from Cannock.
2 ,, Penkridge.

POTTAL POOL.
3 from Cannock.
3 ,, Penkridge.
1 ,, Military Cemetery.

PYE GREEN.
2¾ from Cannock.
1½ ,, Hednesford.

RADMORE.
½ from Castle Ring.

RAWNPIKE OAK.
2½ from Hednesford.
½ ,, Castle Ring.

SEVEN SPRINGS.
¾ from Colwich.
1¼ ,, Wolseley Bridge.

SHARESHILL CHURCH
3 from Cannock.
6 ,, Wolverhampton.

SHOAL HILL.
1 from Cannock.

SHERBROOKE VALLEY.
4½ walk from Military Cemetery through the Valley to emerge on main road from Milford to Rugeley at a point 1½ from former and 4 from latter.

SHUGBOROUGH HALL
2 from Milford.

SOUTH STREET.
4½ walk from Military Cemetery to junction of Stafford-Rugeley main road near Wolseley Bridge.

STILE COP.
2 from Rugeley.
4 ,, Hednesford.

TEDDESLEY HALL.
4½ from Cannock via Shoal Hill.
1¾ ,, Penkridge.
5 ,, Stafford.

WATLING STREET in Chase area.
8 in extent from Gailey to Brownhills.

WEDGES MILLS.
1 from Cannock.

WOLSELEY BRIDGE and
WOLSELEY HALL.
2 from Rugeley.
3½ ,, Milford.

WYRLEY GROVE.
1½ from Norton Canes.
3½ ,, Cannock.

Lightning Source UK Ltd.
Milton Keynes UK
UKHW010700210820
368607UK00003B/617

9 781528 702430